Basic Electronics is published in five parts

Book A Introducing Electronics
ISBN 0 340 41495 2

Book B Resistors, Capacitors and Inductors
ISBN 0 340 41494 4

Book C Diodes and Transistors
ISBN 0 340 41493 6

Book D Analogue Systems
ISBN 0 340 41492 8

Book E Digital Systems
ISBN 0 340 41491 X

It is also available as one complete volume:
ISBN 0 340 41490 1

Note about the author
Malcolm Plant is a Principal Lecturer in the Faculty of Education at Nottingham Polytechnic. He is the author of several books, including *Teach Yourself Electronics* (Hodder & Stoughton 1988). His main professional interests are in astronomy and astrophysics, electronics instrumentation and issues relating to conservation and the environment.

Contents

Resistors

1 What Resistors Do

1.1 Reminder **1**
1.2 The meaning of electrical resistance **1**
1.3 Resistance and current flow **1**

2 Coulombs, Amperes and Volts

2.1 Electric charge and the coulomb **3**
2.2 Electric current and the ampere **3**
2.3 What the volt measures **4**
2.4 *Experiment B1*:
Working out the resistance of a filament lamp **5**

3 Resistor Types, Values and Codes

3.1 Introduction **7**
3.2 The wire-wound resistor **7**
3.3 The moulded carbon resistor **7**
3.4 Carbon-film, metal-film, metal-oxide and thick film resistors **8**
3.5 Variable resistors **8**
3.6 Resistor values and codings **9**
3.7 An alternative code: the BS 1852 code **11**
3.8 Preferred values of resistors **11**
3.9 *Experiment B2*:
Using an analogue multimeter to measure resistance **12**

4 Ohm's Law

4.1 Introduction **14**
4.2 Calculations **14**
4.3 Resistors in series **15**
4.4 Resistors in parallel **16**
4.5 More calculations **17**
4.6 Cells in series and parallel **18**
4.7 Internal resistance **19**

5 The Potential Divider and the Wheatstone Bridge

5.1 The potential divider **20**
5.2 Loading a potential divider **21**
5.3 The Wheatstone bridge **22**

6 Electrical Energy and Power

6.1 Introduction **24**
6.2 The joule **24**
6.3 The watt **25**
6.4 Calculating electrical power **25**
6.5 Power ratings of resistors **26**
6.6 Using a light-emitting diode **27**
6.7 Proving that power = volts × amps **28**

7 Special Types of Resistor

7.1 Introduction **30**
7.2 What a light dependent resistor looks like **30**
7.3 *Experiment B3*:
Measuring the resistance of an LDR **30**
7.4 *Experiment B4*:
Using an LDR **31**
7.5 *Experiment B5*:
Making a simple lightmeter **32**
7.6 What a thermistor looks like **32**
7.7 *Experiment B6*:
Measuring the resistance of a thermistor **32**
7.8 *Experiment B7*:
Using a thermistor in a voltage divider **34**
7.9 *Experiment B8*:
Designing a simple thermometer **34**
7.10 What a strain gauge looks like **35**
7.11 Gauge factor **36**
7.12 Temperature effects **37**
7.13 Why measure strain anyway? **38**

621.381

BASIC ELECTRONICS

Resistors, Capacitors and Inductors

Malcolm Plant

Hodder & Stoughton
LONDON SYDNEY AUCKLAND TORONTO

Learning without thought is labour lost;
thought without learning is perilous.

Confucius, *Analects*

Every effort has been made to trace and acknowledge ownership of
copyright. The publishers will be glad to make suitable
arrangements with any copyright holders whom it has not been
possible to contact.

© 1990 SCDC Publications

First published in Great Britain 1976

Second edition 1990

British Library Cataloguing in Publication Data.
Plant, M. (Malcolm), 1936–
    Basic electronics. — 2nd ed.
    Book B. Resistors
    1. Electronic equipment. For schools
    I. Title
    621.381
    ISBN   0 340 41494 4
    ISBN   0 340 41490 1 set

Typeset in 11/12 Baskerville by Taurus Graphics, Abingdon, Oxon.
Printed and bound in Great Britain for the educational publishing
division of Hodder and Stoughton, Mill Road, Dunton Green,
Sevenoaks, Kent by Thomson Litho Ltd, East Kilbride.

Capacitors

8 What Capacitors Do

8.1 Capacitors store electric charge **39**
8.2 Units of capacitance **39**
8.3 The symbol and structure of a capacitor **39**
8.4 Operating characteristics of a capacitor **40**
8.5 Types of capacitor **41**

9 Basic Experiments with Capacitors

9.1 *Experiment B9*:
Testing a capacitor with a multimeter **45**
9.2 *Experiment B10*:
Charging a capacitor **45**
9.3 *Experiment B11*:
Storing energy in a capacitor **47**

10 Time Constant

10.1 Introduction **48**
10.2 *Experiment B12*:
Measuring time constant **48**
10.3 *Experiment B13*:
Discharging a capacitor **49**
10.4 The importance of time constant **50**
10.5 Charge and discharge equations **50**
10.6 Important notes **52**

11 Combinations of Capacitors

11.1 Introduction **53**
11.2 *Experiment B14*:
Connecting capacitors in parallel **53**
11.3 The formula for capacitors connected in parallel **54**
11.4 *Experiment B15*:
Connecting capacitors in series **54**
11.5 The formula for capacitors connected in series **54**
11.6 Examples **54**

12 Energy Stored in a Capacitor

12.1 Introduction **56**
12.2 The charge stored in a capacitor **56**
12.3 The energy stored in a capacitor **57**

13 Capacitors in Alternating Current Circuits

13.1 Introduction **58**
13.2 *Experiment B16*:
Showing that a capacitor passes a.c. **58**
13.3 *Experiment B17*:
Measuring the reactance of a capacitor **59**
13.4 Calculating reactance values **60**
13.5 Coupling and decoupling capacitors **60**

14 The 555 Timer

14.1 Introduction **62**
14.2 *Experiment B18*:
Designing a monostable using the 555 timer **62**
14.3 *Experiment B19*:
Designing an astable using the 555 timer **64**
14.4 *Experiment B20*:
Designing a timer/alarm **66**

Inductors

15 What Inductors Do and How They Are Made

15.1 What an inductor does **67**
15.2 How an inductor is made **67**
15.3 *Experiment B21*:
Showing the effect of a conductor **68**

16 The Magnetic Effects of a Current

16.1 Introduction **69**
16.2 *Experiment B22*:
Showing the magnetic effect of a current **69**
16.3 *Experiment B23*:
Investigating the magnetic field produced by a solenoid **70**
16.4 *Experiment B24*:
Making a buzzer **71**
16.5 Loudspeakers **72**

17 How Magnetism Produces Electricity

17.1 Introduction **74**
17.2 *Experiment B25*:
Producing electricity from magnetism **74**
17.3 *Experiment B26*:
Showing the effect of self-inductance **74**
17.4 Units for measuring inductance **76**

18   Time Constant of an Inductive Circuit

18.1   Reminder   **77**
18.2   Increasing current in an *L–R* series circuit   **77**
18.3   Decreasing current in an *L–R* series circuit   **77**
18.4   *Experiment B27*: Showing the effect of inductive reactance   **77**

19   Transformers

19.1   What a transformer does   **80**
19.2   How a transformer works   **80**
19.3   Why a transformer core is laminated   **81**
19.4   The transformer equation   **82**
19.5   Inductor and transformer core materials   **83**

20   Electromagnetic Relays

20.1   What a relay does   **85**
20.2   The structure of a simple relay   **85**
20.3   The relay symbol and contacts   **85**

20.4   Using a protective diode in relay circuits   **86**
20.5   *Experiment B28*: Detecting back e.m.f.   **87**

21   The Tuned Circuit

21.1   Resonance   **88**
21.2   Selectivity   **89**

22   Project Modules

22.1   What they are   **90**
22.2   *Project Module B1*: Schmitt Trigger   **91**
22.3   *Project Module B2*: Relay Driver   **94**
22.4   *Project Module B3*: Bistable   **96**
22.5   *Project Module B4*: Monostable   **98**
22.6   *Project Module B5*: Radio Receiver   **101**
22.7   *Project Module B6*: Infrared Source and Sensor   **105**
22.8   *Project Module B7*: Metal Detector   **108**

23   Questions and Answers

Revision questions, Revision answers, Answers to questions

Summary

Resistors, whether they are discrete (i.e. separate) or integrated (i.e. formed with other components on a silicon chip) are the most commonly used components in electronic circuits. Because they offer an electrical resistance to the flow of current, resistors are used to adjust the voltage at various points in a circuit which ensures that devices such as transistors and integrated circuits operate correctly. Chapters 1 to 6 describe various types of resistor, the resistor colour code, the use of resistors in voltage dividers, Ohm's law, series and parallel connections of resistors and electrical energy and power.

Chapter 7 introduces the basic properties of three widely used special types of resistor, the light-dependent resistor (LDR), the thermistor and the strain gauge. The LDR enables light-operated circuits to be designed which switch on or off at predetermined light levels. The thermistor enables temperature control and thermometer circuits to be designed and the strain gauge can be used in circuits for measuring how much mechanical strain, or bending, a mechanical object undergoes when it is loaded. As well as having the few basic applications described in Chapter 7, these three special types of resistor are used in a wide range of circuits throughout *Basic Electronics*.

Capacitors are almost as common as resistors in circuits, and are used in many applications including oscillators, radio receivers and transmitters, filters and timers. Capacitors owe their usefulness to the fact that they can store electrical charge, as the simple experiments in Chapters 8 and 9 show.

A capacitor stores charge when a current flows through a resistor connected in series with it. Both the value of the capacitor and that of the resistor determine the rate at which charge is stored, a property which leads to the concept of 'time constant' as explained in Chapter 10. As you will see, the time constant enables you to predict the time delay you expect when designing timers and oscillators. Like resistors, capacitors can be connected in series and parallel to provide different values of capacitance as described in Chapter 11. Chapter 12 enables you to work out how much electrical energy you can expect to store in a capacitor.

Though capacitors do not allow direct current to pass through them, they do pass alternating current signals. This makes a capacitor a very useful component in audio amplifiers and filters, for example. The effective resistance, or 'capacitive reactance', of capacitors to a.c. signals depends on the frequency of the signals as explained in Chapter 13.

There are a number of integrated circuits which rely on the use of resistors and capacitors to produce timers and oscillators and the 555 timer IC is the best known of these. Chapter 14 introduces this useful IC and explains how you can assemble astable and monostable circuits using it.

Chapter 15 explains that inductors are basically coils of wire. They have the property of 'inductance' which means they oppose changing currents passing through them as explained in Chapters 16 and 17. This opposition to a changing current is discussed in Chapter 18 and is called inductive reactance.

The transformer is a device for converting one a.c. voltage to another. It works by ensuring that the changing magnetic field of one coil passes through another coil which has a different number of turns of wire. The properties and uses of transformers are explained in Chapter 19.

Another useful device which was briefly

introduced in Section 5.2 of Book A is the electromagnetic relay. You will find a more detailed description of the operation of the electromagnetic relay in Chapter 20.

Chapter 22 contains a description of seven Project Modules. These are units for the rapid assembly of electronic systems. They can be built from the printed circuit board (PCB) design provided. You should turn to Chapter 12 in Book A for an introduction to their use.

As with every book of *Basic Electronics*, Book B ends with Answers to Questions in the text, Revision Questions and Revision Answers.

If you want to follow an easier and quicker route through Book B, you should omit the sections of the text marked with the symbols ∇ and $\triangle$ in the left hand margin.

1 What Resistors Do

1.1 Reminder

In Section 4.5 of Book A you were able to assemble a simple circuit consisting of a lamp and a battery. The battery supplied the electrical energy which the filament of the lamp converted into light energy (and heat). It was explained that electrons, flowing from the negative to the positive terminal of the battery, do work in passing through the filament, and this work appears in the form of light energy emitted from the filament.

Questions

1 Draw a circuit diagram consisting of a battery and a lamp connected in series with each other. Mark on your circuit suitable values for the electromotive force (e.m.f.) of the battery and the voltage-rating of the lamp.
2 Show the direction in which electrons move round the circuit you have drawn.
3 What is meant by 'series' and 'parallel' connection of two lamps in a circuit?

1.2 The meaning of electrical resistance

When a component in a circuit converts electrical energy into heat, we say that the component has *electrical resistance*. The component resists the flow of electrons through it and, as you have seen with a lamp filament, causes electrical energy to be converted into heat. We often say electrical energy is 'consumed' in a component which has resistance; what we really mean is that electrical energy is converted into some other form of energy, in this case heat energy. It is never possible

to make energy disappear, for it always appears in some other form.

Electrical resistance is a nuisance in a circuit if the heat produced interferes with the operation of other components nearby. You will be shown how to calculate the resistance of a component as well as calculating the electrical energy 'consumed' by it.

Question

1 Can you think of examples where electrical resistance is useful?

1.3 Resistance and current flow

The circuit of figure 1.1 can be used to get a rough idea of the effect of resistance in a circuit. This circuit should be familiar to you, since it was used in Section 4.3 in Book A as an example of a series circuit. Since the lamps are identical and the same current flows through them, they glow equally brightly. In other words, the filaments have resistances which are equal. This is shown by figure 1.1(a), overleaf.

The wire W can be used to 'short out' L_1 by making contact at P. This wire is made of copper, which has a very low resistance. Therefore, when L_1 is shorted out, current prefers to flow through the wire rather than through the lamp L_1. L_1 goes out, and at the same time L_2 glows more brightly as shown in figure 1.1(b), overleaf.

Now we know why L_1 goes out, but why does L_2 glow more brightly? The resistance of L_2 is unaltered and so is the e.m.f. of the battery, but since the resistance of the circuit has been halved, more current flows

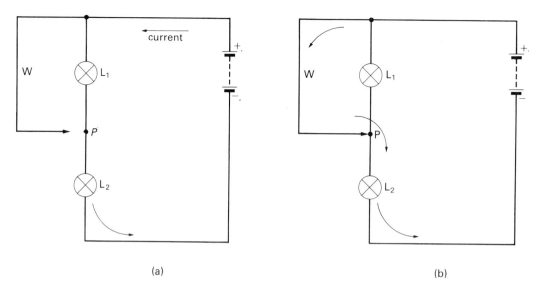

Figure 1.1 Seeing how resistance affects the brightness of a lamp

through lamp L_2 so that more energy is converted into heat and light in it. Once you understand how resistance and current flow are related, a more precise explanation can be given.

2 Coulombs, Amperes and Volts

2.1 Electric charge and the coulomb

You know that an electric current consists of a flow of electrons. These electrons are negatively charged (see Chapter 7 of Book A) and are an important part of an atom. The *coulomb* (pronounced 'coo-lom') is used to measure *electric charge*.

The fact that about six million million million electrons are required to provide a total charge of one coulomb should give you an idea of how small is the charge carried by an electron!

Remember

When electrons move, electric charge is moving; and when electric charge moves we have an electric current.

2.2 Electric current and the ampere

You have already used a filament lamp rated at 6 V, 0.06 A (0.06 A = 60 mA).

Question

1 What do the letters V and A stand for?

The ampere is used to indicate the strength of a current, and is measured in units of *coulombs per second*. In other words, electric current is the *rate* at which electric charge in coulombs moves through a circuit. It is easy to remember that one ampere is electric charge flowing at the rate of one coulomb per second, or

$$1A = 1 \text{ coulomb/second} = 1 \text{ C/s}.$$

So how many coulombs per second flow through the filament of a 6 V, 60 mA lamp? Obviously, if 1 A is 1 coulomb per second, 60 mA = 0.06 coulombs per second.

Questions

2 How many coulombs pass through a car headlamp in one second if it is rated at 12 V, 4 A?

3 A current of 300 milliamps (mA) flows through a lamp. How many coulombs flow through it per second?

The letter I is the symbol used for electric current. Thus we can write $I = 0.06$ A for the current flowing through the filament lamp. There is one important point about current flowing in a parallel circuit like figure 2.1.

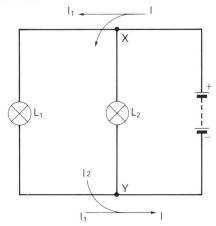

Figure 2.1 Current flowing in a parallel circuit

The current I from the battery divides at point X, part I_1 flowing through L_1, and part I_2 flowing through L_2. I_1 and I_2 add together at point Y to give I again.

Thus, $I = I_1 + I_2$.

4   What can you say about I_1 and I_2 if the lamps are identical?

Remember

Current is measured in amperes and it flows through a component from one terminal to another.

An ammeter measures current and it is connected in series with the component through which the current flows (see Section 9.2 in Book A).

2.3   **What the volt measures**

As you know from Book A, the e.m.f. of a battery is the electrical force between its terminals which makes current flow round a circuit connected to it. The volt is the unit for measuring e.m.f.: for example, we say 'the e.m.f. of a battery is 9 volts.'

We call the electrical force which makes current flow through a component, e.g. a resistor, a potential difference (p.d.). It is important to be able to calculate the p.d. across components in a circuit to make sure the circuit is being operated correctly.

Remember

Electromotive force (e.m.f.) is the electrical force, measured in volts, between the terminals of a battery which makes current flow through a circuit connected to it.

Potential difference (p.d.) is the difference of electrical force, measured in volts, between the terminals of a component which makes current flow

through the component. The volt is the unit of e.m.f. and p.d. and it is measured using a voltmeter which is connected in parallel with the component through which the current flows.

Questions

1   What is wrong with the circuit shown in figure 2.2?

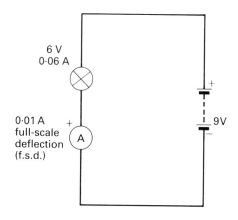

Figure 2.2   What is wrong with this circuit?

2   Name three errors in the circuit shown in figure 2.3, which was set up to find the current flowing through the lamp.

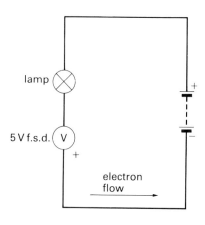

Figure 2.3   Name three errors in this circuit

3 In the circuit of figure 2.4, you want to measure the total current flowing from the battery. Where would you insert the ammeter — at point X, Y or Z?

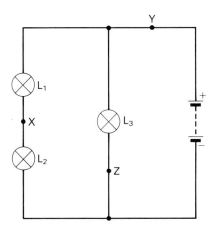

Figure 2.4 Where do you put an ammeter to measure the total current flowing in the circuit?

4 In the circuit of figure 2.4, you want to measure the voltage across lamp L_2. To which two of the points X, Y and Z do you connect the voltmeter?

2.4 *Experiment* B1

Working out the resistance of a filament lamp

The circuit shown in figure 2.5 shows how an ammeter and a voltmeter are used to measure the resistance of the filament of the lamp, L_1. Figure 2.6 overleaf shows how a length of terminal block can be used to assemble the circuit required.

If you have two multimeters, you will be able to measure the current and the voltage at the same time.

Suppose you find: current = 50 mA = I
voltage = 6 V = V

Find the ratio V/I.

$$\frac{V}{I} = \frac{6 \text{ V}}{50 \text{ mA}} = \frac{6}{0.050} = 120$$

This number has the units of *ohms*, and is the electrical resistance of the filament. It has the Greek symbol Ω (omega).

Remember

When the voltage across a component is divided by the current flowing through the component, the result is the resistance of that component measured in ohms. The component in your experiment is the filament of the lamp.

Resistance (ohms) = volts/amperes

or $R = \dfrac{V}{I}$

This simple equation is very helpful when designing electronic circuits. Its use is more fully explained in Chapter 4.

Questions

1 A car headlamp bulb is rated at 12 V, 4 A. What is its resistance?
2 A domestic lamp is rated at 240 V and carries a current of 0.25 A. What is its resistance?
3 A filament bulb is rated at 6 V, 60 mA. What is its resistance?
4 What is the origin of each of these four words: *coulomb, ampere, volt* and *ohm*?

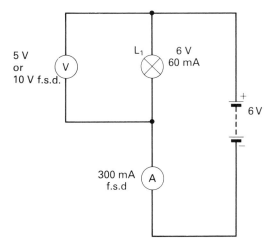

Figure 2.5 Circuit for measuring the resistance of the filament of a lamp

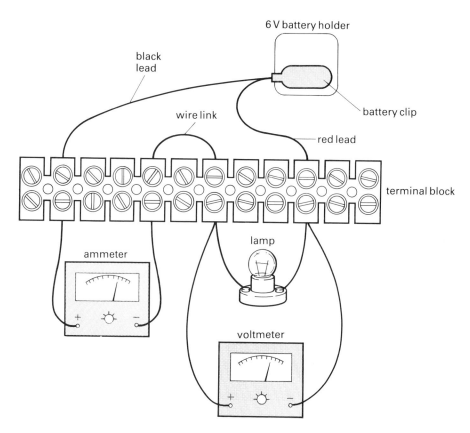

Figure 2.6 Experiment B1: Working out the resistance of a filament lamp

3 Resistor Types, Values and Codes

3.1 Introduction

Components which have an electrical resistance are called resistors. Two types of resistor are in common use, the fixed-value resistor and the variable resistor. Resistors are used to control the flow of current: so that the correct voltages are obtained at various points in a circuit as explained in Chapter 5.

3.2 The wire-wound resistor

An alloy of the metals nickel and chromium is called *nichrome*. This alloy has a higher electrical resistance than copper wire, and is used for electric-fire bars and for some types of resistor. The alloy constantan (or eureka) is also used for wire-wound resistors.

Figure 3.1 shows a fixed-value resistor made of resistance wire. There are three factors which determine the electrical resistance of wire-wound resistors:

(a) the type of wire used
(b) the thickness of the wire
(c) the length of wire on the former

Wire-wound resistors generally have values in the range from fractions of an ohm to about 25 000 ohms (25 kΩ). They can be made with great precision and readily dissipate heat produced by them.

3.3 The moulded-carbon resistor

The structure of this type of resistor is shown in figure 3.2. Values range from a few ohms to ten million ohms (10 MΩ). It is suitable for all general-purpose switching circuits but not for audio circuits because it is 'electrically noisy' and causes hiss in the sound ouput from amplifiers.

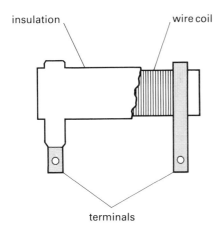

Figure 3.1 The parts of a wire-wound resistor

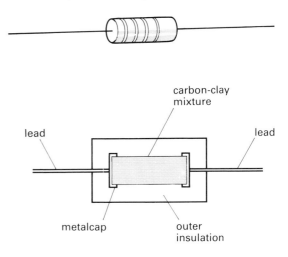

Figure 3.2 A moulded-carbon resistor

3.4 Carbon-film, metal-film, metal-oxide and thick-film resistors

Carbon-film, metal-film and metal-oxide resistors are able to perform reliably for long periods in conditions of wide temperature and humidity changes. The carbon-film types are made by depositing a film of carbon on a ceramic rod which is then protected by a hard-wearing electrically-insulating coating. The metal-film resistor has a tin oxide film instead of a carbon film. These resistors are recommended for use in audio circuits, e.g. amplifiers and radio receivers, and where circuits are exposed to extreme changes of temperature and humidity.

Two examples of thick-film resistors are shown in figure 3.3. These resistors are made by adjusting the thickness of a layer of semiconducting material to give the required resistance. The resistors are packaged either in a single-in-line (SIL) package or a dual-in-line (DIL) package. The resistors may be independent of each other, or have a common connection depending on the application. For example, the DIL package is very useful in seven-segment display circuits (see Book E) where each segment of a seven-segment display needs a resistor in series with it. Such a package is also useful in microcomputer circuits where a number (usually eight) of data lines have to be *interfaced* to control and display circuits.

3.5 Variable resistors

The variable resistor is widely used as a volume control in radios and audio circuits, and it can also be used to control the brightness of a lamp or the speed of a motor. In fact, the variable resistor is properly called a *potentiometer* (or pot, for short).

A typical spindle potentiometer and the symbol used for it in circuit diagrams is

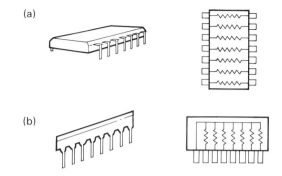

Figure 3.3 Two types of package for thick-film resistors. Part (a) shows a DIL package, Part (b) shows a SIL package.

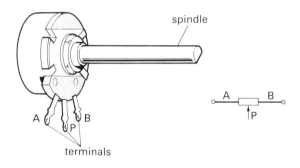

Figure 3.4 A potentiometer or variable resistor

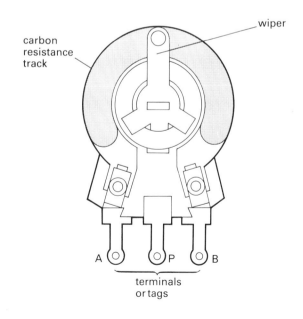

Figure 3.5 The structure of a potentiometer

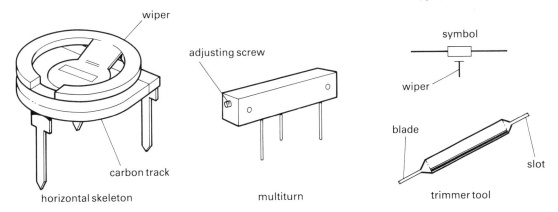

Figure 3.6 Two types of preset potentiometers

shown in figure 3.4. The terminal P is the *wiper* which moves round a fixed circular track of resistance wire or carbon when the spindle is operated. As figure 3.5 shows, if the wiper moves along the track from A to B, the resistance between P and A increases, and between P and B decreases. Thus, by using just the slider and terminal A or B, a variable resistance can be obtained. However, it is useful in some circuits to use all three terminals so that the potentiometer acts as a potential divider — see Chapter 5. Note that terminals A, B and P have holes in them so that wires can be inserted through the holes and soldered to the terminals.

There are two types of resistance variation as the spindle is rotated. In a *linear track* potentiometer, equal changes of resistance occur when the potentiometer is rotated through equal angles. In a *log track* potentiometer, the resistance towards one end of the track progressively decreases for equal angles of rotation. The resistance of spindle potentiometers (i.e. between the ends of their tracks) ranges from 1 kΩ to 1 MΩ; lower and higher values are less often used.

Figure 3.6 shows two types of miniature *preset* potentiometers and the symbol used for them in circuit diagrams. Presets have three terminals and are intended to be soldered directly in a circuit board. A small trimmer tool enables the wiper to be precisely adjusted along the carbon track to make fine adjustments to the performance of a circuit. A multiturn preset requires 10

or 20 turns to move the wiper between the two ends of the track. Preset potentiometers are not intended to be adjusted frequently.

3.6 **Resistor values and codings**

In electronic circuits small-value currents flow through large-value resistors. As explained in Book A, a current of one milliampere (1 mA) is one thousandth of an ampere. A current of one microampere (1 μA) is one millionth of an ampere. But resistor values are rarely less than 1 ohm and circuits often use resistors measured in thousands and millions of ohms.

Remember

1 thousand ohms = 1 kilohm or 1 kΩ, and 1 million ohms = 1 megohm = 1 MΩ

1 thousandth of an ampere = 1 milliampere or 1 mA
1 millionth of an ampere = 1 microampere or 1 μA

The relationship between volts and amperes and ohms is described in Chapter 4.

Some typical resistor values are:

1.2 kΩ (1200 ohms)
47 kΩ (47 000 ohms)
330 kΩ (330 000 ohms)
2.2 MΩ (2 200 000 ohms)

Questions

1   What short way is there to write:
    (a)   12 000 ohms?
    (b)   220 thousand ohms?
    (c)   3 million ohms?
2   Write 0.1 MΩ in kΩ.
3   Write 470 Ω in kΩ.
4   How many kΩ s are there in 1 MΩ?

Most fixed-value resistors are marked with coloured bands so that their values can be read easily.

Questions

5   What advantage has the use of coloured bands rather than letters and numbers?
6   Who would find the colour-code difficult to interpret?

Figure 3.7 lists the colours used in the *resistor colour-code* and the values attached to the coloured bands.

Band 1 gives the first digit of the value.
Band 2 gives the second digit of the value.
Band 3 gives the number of zeros which follow the first two digits.
Band 4 gives the *tolerance* of the value worked out from the first three bands.

Example 1

Band	1	2	3	4
Colour	red	violet	brown	silver
Value	2	7	0	10%

Therefore, this resistor has a value of 270 Ω to within 10%, more or less. That is, the resistor might have a value between 270 Ω + (10% of 270) Ω and 270 Ω − (10% of 270) Ω, or between (270 + 27) Ω and (270 − 27) Ω. Thus the value lies somewhere between 243 and 297 ohms.

Example 2

Band	1	2	3	4
Colour	yellow	violet	red	gold
Value	4	7	00	5%

This resistor has a value of 4700 ohms to within 5% more or less. That is, its value is 4.7 kΩ ± 5%.

Figure 3.7 The resistor colour-code

colour	band 1	band 2	band 3	band 4
black	0	0	none	1%
brown	1	1	0	2%
red	2	2	00	3%
orange	.3	3	000	4%
yellow	4	4	0 000	–
green	5	5	00 000	–
blue	6	6	000 000	–
violet	7	7	0 000 000	–
grey	8	8	–	–
white	9	9	–	5%
gold	–	–	0·1	10%
silver	–	–	0·01	20%
no colour	–	–	–	

Questions

7   Three resistors are colour-coded as follows:

Band	1	2	3	4
Resistor A	green	blue	orange	silver
Resistor B	red	red	red	red
Resistor C	orange	orange	black	gold

What are the values and tolerance of these resistors?

8   You are looking for a 5.6 kΩ resistor. What is the colour of the third band?

9   What would be the colour coding on a 1.5 MΩ resistor of 5% tolerance?

In order to give higher precision in the value of resistors, some manufacturers are using a five-band colour code. The first three bands give the first three digits of the value and the fourth band gives the number of zeros. The fifth band gives the tolerance. The extra digit band means that a resistor can have a value specified to, say, 1.27 kΩ.

For most circuits, four-band resistors having a tolerance of 10% or 5% are quite satisfactory. It is unnecessary to use resistors accurate to, say, 1.27 kΩ.

▽ 3.7 **An alternative code: the BS 1852 code**

This code consists of letters and numbers instead of colours, and is used on variable as well as fixed-value resistors. The following are examples of this British Standards (BS) code:

5R6J = 5.6Ω ± 5%
2K2K = 2.2kΩ ± 10%
47KK = 47 kΩ ± 10%
2M2G = 2.2 MΩ ± 2%

In this code, the letters R, K and M are multiplier factors when they are the first of the two letters in the code. And the position of this letter indicates the decimal point.

R is times 1; K is times 1000;
M is times 1 000 000

The second letter of the two gives the tolerance:

F = 1%; G = 2%; H = 2.5%;
J = 5%; K = 10%; M = 20%

Look at the first example — 5R6J:

5	R	6	J
1st digit	decimal point and × 1 factor	second digit	tolerance 5%

Thus this is a 5.6 Ω resistor with a 5% tolerance. Look carefully at the other examples, and check that you understand the code.

Questions

1   What is the value and tolerance of each of the following resistors marked with the BS1852 code:

    (a)   1M2F?
    (b)   150RM?
    (c)   12KJ?
    (d)   68KK?

2   Write BS 1852 codes for the following resistors:

    (a)   330 kΩ ± 10%,
    (b)   47 Ω ± 5%.

▽ 3.8 **Preferred values for resistors**

When manufacturers produce a 100 Ω resistor with a tolerance of 10%, there is no point in them making one of 105 Ω with the same tolerance. Their best plan is to ensure that the next highest resistor made has its lowest possible value equal to the highest possible value of the 100 Ω resistor.

Question

1   What is the upper value of the 100 Ω resistor?

Suppose a manufacturer provides a 120 Ω resistor as his next value above, and it has a tolerance of 10%. Its lowest possible value is 120 — (120/10) = 108 Ω, which is below the highest possible value of 110 Ω for the 100 Ω ± 10% resistor.

The resistor values which fulfil these conditions of overlap are called *preferred values*, and manufacturers produce only these values (except for special resistors such as wire-wound types). For 10% tolerance resistors, the preferred values belong to the so-called 'E12 series'. The twelve values are 10, 12, 15, 18, 22, 27, 33, 39, 47, 56, 68, and 82.

These digits are the first two in the colour code, and are the only ones which need remembering. The third band, of course, gives the number of zeros to be added to these figures.

Note that there is no point in looking for a four-band resistor in the E12 series with a value of 123 ohms — there is usually no need for this precision anyway. You will have to settle for a nominal 120 ohm resistor which might have a value of 123 ohms.

Resistors belonging to the E24 series have a 5% tolerance and 24 different values. In addition to the values in the E12 series, the extra preferred values include the factors 1.1, 1.3, 1.6, 2.0, 2.4, 3.0, 3.6, 4.3, 5.1, 6.2, 7.5 and 9.1.

Questions

2   What preferred value resistor in the E12 series would you look for if a calculation showed the following values were needed:

(a)   105 Ω?
(b)   35 Ω?
(c)   700 Ω?
(d)   1 428 390 Ω?

3   Prove that a 3.3 kΩ ± 10% resistor could have an upper value which is greater than the lower value of a
△   3.9 kΩ ± 10% resistor.

3.9 *Experiment* B2

Using an analogue multimeter to measure resistance

If you have used an analogue multimeter (see Chapter 9, Book A), you will notice that the resistance scale starts from the right-hand end of the scale (zero ohms is here) and the scale is non-linear, being cramped towards the left-hand end of the scale (at high resistance values) — see figure 3.8.

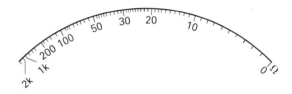

Figure 3.8   The ohms scale of an analogue multimeter

The multimeter is able to measure resistance because it has an internal battery. When the range switch is operated to measure ohms, this internal battery sends current through a resistor connected between the leads.

Although there are small variations in multimeters, the following instructions will enable you to check the value of, say, a 4.7 kΩ resistor.

(a)   Switch the multimeter to read 'ohms × 1000' or 'ohms × 1k'.
(b)   Connect the multimeter leads together. The needle will swing towards the zero (right-hand end) of the scale but it will not necessarily come to rest exactly on zero.
(c)   Operate the 'ohms adjust' control and 'zero' the meter.

(d)   Connect the leads across the ends of the resistor, as shown in figure 3.9, and read off the resistance. The meter is acting as an *ohmmeter*.

Questions

1   Does the value you measure lie within the range of possible values for the resistor? See Section 3.8.
2   Repeat your measurements with other resistors, both high and low values.

Remember

Each time you switch to a new range of resistance, zero the meter as in **(b)**. This is necessary as the e.m.f. of the internal battery falls during its working life.

Always switch the multimeter to a current or voltage range, or to its off position, after you have used it to measure ohms. Why do you think this is necessary?

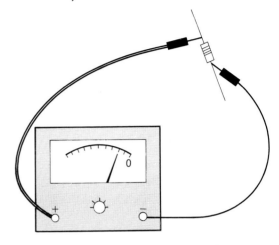

Figure 3.9 Experiment B2: Using an analogue multimeter to measure resistance

4 Ohm's Law

4.1 Introduction

The experiment in Section 2.4 explains how the resistance of a filament lamp might be measured. A voltmeter is used to measure the potential difference (V) across the lamp and an ammeter the current (I) through the lamp. The resistance R of the lamp is given by $R = V/I$.

This equation is very useful in designing electronic circuits but it is often wrongly called *Ohm's law*. This important law is:

> Provided the temperature and other physical conditions of an electrical conductor do not change, the potential difference across it is proportional to the current flowing through it.

'Proportional' means that doubling the p.d. (voltage) doubles the current, and halving the p.d. halves the current, and so on. In other words, Ohm's law means that the ratio V/I does not vary. And since $V/I = R$, the resistance of the conductor does not vary as its physical state does not change.

In electronics, we often make use of the equation $R = V/I$. All the equation tells us is the relationship between resistance, p.d. and current for a conductor, and this relationship is true whether Ohm's law is true or not. In Section 2.4 this equation was used to calculate the resistance of the filament lamp and Ohm's law is certainly not true for the heated tungsten filament of the lamp. In fact, the resistance increases with the brightness of the lamp, i.e. with the temperature of the tungsten.

4.2 Calculations

As well as wanting to calculate the resistance R from a knowledge of V and I, you will also need to calculate V and I. The

Figure 4.1 A triangle for remembering the relation between V, I and R

triangle of figure 4.1 is designed to help you remember how to find R, V, and I.

To find R, cover R to give $R = V/I$.
To find V, cover V to give $V = I \times R$.
To find I, cover I to give $I = V/R$.

Example 1

A filament lamp passes a current of 60 mA (0.06 A) when the voltage across it is 6 V. What is the resistance?

$R = V/I = 6/0.06 = 100$ ohms

Example 2

What is the current flowing through a 30 ohm wire-wound resistor when it is connected to a 9 V battery?

$I = V/R = 9/30 = 0.3$ A or 300 mA

Example 3

What is the voltage across the ends of a 2.2 kΩ resistor when 4 mA flows through it?

$$V = I \times R = 2.2 \text{ k}\Omega \times 4 \text{ mA}$$
$$= 2.2 \times 1000 \times (4/1000)$$
$$= 8.8 \text{ V}$$

Now use one of the equations $R = V/I$, $V = IR$, or $I = V/R$ to answer the following questions:

Questions

1   A car flasher bulb is rated at 12 V, 1.75 A. What is its resistance when alight?

2   The voltage across a 5.0 kΩ resistor is 6 V. What current flows through it?

3   A current of 1.3 mA flows through a 10 kΩ resistor. What is the voltage across the ends of the resistor?

4   A special kind of resistor known as a *thermistor* — see Section 7.6 — allows a current of 4.5 mA to pass through it when the voltage across it is 9 V. When the thermistor is heated, the current rises to 6 mA. Has its resistance increased or decreased? What is the change of resistance?

4.3   Resistors in series

Resistors are very common components in electronic circuits, and they are sometimes connected together in particular ways to produce resistance values not easily obtained with single resistors.

Figure 4.2 shows how to connect two resistors in series. The milliammeter measures the current which flows through R_1 and R_2.

Questions

1   Does the same current flow through R_1 as flows through R_2? How would you check this?

2   What can you say about the potential difference across each of the resistors?

There are two important points to remember about resistors in series:

(a)   The same current flows through each resistor.

(b)   The sum of the p.d.s. across each resistor is equal to the p.d. across the combination.

Thus in figure 4.3, $V = V_1 + V_2$: and, since $V = IR$, we can write:

$$V = IR_1 + IR_2 = I(R_1 + R_2) = IR$$

Here we have written R for $R_1 + R_2$. So if we replace R_1 and R_2 by a single resistor whose value is $(R_1 + R_2)$, the current drawn from the battery remains unaltered.

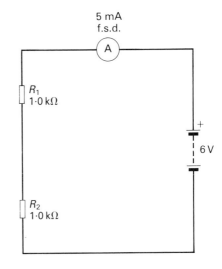

Figure 4.2   A series resistor-test circuit

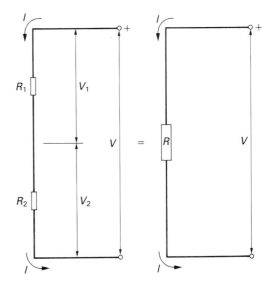

Figure 4.3   The current is the same if $R = R_1 + R_2$

Remember

The equivalent resistance of two resistors in series is equal to their sum; that is, $R = R_1 + R_2$. Note that the combined resistance is higher than either resistance.

It should be clear that the single resistor required to replace three or more resistors connected in series has a value equal to their separate values added together.

Questions

3   What is the value of the single resistor which would replace R_1 and R_2 in figure 4.2 and leave the current unaltered?

4   A 4.7 kΩ and a 2.2 kΩ resistor connected in series are to be replaced by a single resistor having the same resistance as the combination. What resistor is required? What is the nearest preferred value?

4.4   **Resistors in parallel**

Figure 4.4 shows how to connect two resistors in parallel. The milliammeter measures the sum of the currents which flow through R_1 and R_2 separately.

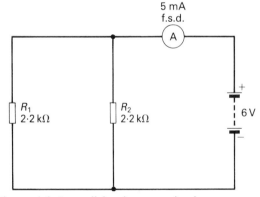

Figure 4.4 A parallel resistor-test circuit

Questions

1   What can you say about the current which flows through each of the resistors?

2   What can you say about the potential difference across each of the resistors?

There are two important facts about resistors in parallel:

(a)   The same p.d. acts across each resistor.

(b)   The sum of the currents through each resistor is equal to the current flowing from the source (battery).

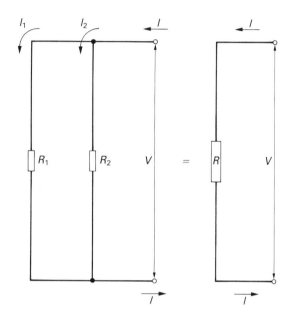

Figure 4.5 The current is the same if $R = R_1 \times R_2/(R_1 + R_2)$

Thus in figure 4.5, $I = I_1 + I_2$; and, since $I = V/R$, we can write:

$$I = V/R_1 + V/R_2 = V/R$$

Since the voltage V is a common factor on both sides of this equation, we can write:

$$1/R = 1/R_1 + 1/R_2$$

or $$R = \frac{R_1 \times R_2}{R_1 + R_2}$$

Remember

The equivalent resistance of two resistors connected in parallel is found by dividing the product of their values by their sum.

To see how to use this equation, let us find the value of the single resistor required to replace the two resistors in the circuit of figure 4.4 so as to leave the current drawn from the battery unaltered.

$$R = \frac{2.2 \times 2.2}{2.2 + 2.2} = 1.1 \text{ k}\Omega$$

Question

3 What single resistor is equivalent to two 4.7 kΩ resistors connected in parallel? What is the nearest preferred value to this resistance?

▽ 4·5 **More calculations**

You want to be able to replace the three resistors shown in figure 4.6 by one single equivalent resistor. The resistor network consists of two parallel-connected resistors (R_2 and R_3) in series with R_1.

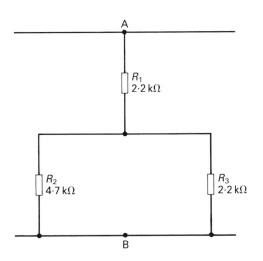

Figure 4.6 A series – parallel resistor combination

First find the equivalent resistance of R_2 and R_3 as follows:

$$R = \frac{4.7 \times 2.2}{2.2 + 4.7}$$
$$= 1.5 \text{ k}\Omega \text{ (approximately)}$$

A resistor of 1.5 kΩ in series with one of 2.2 kΩ can be replaced by a preferred one of 3.6 kΩ since actual resistors have a tolerance of, say, ± 5%. Therefore, a 3.6 kΩ resistor connected across AB will have the same resistance as the network between AB.

Questions

1 What is the equivalent resistance of three resistors of values 1.2 kΩ, 2.2 kΩ, and 4.7 kΩ connected in series?
2 What is the equivalent resistance of a 330 kΩ and a 100 kΩ resistor connected in parallel?
3 Calculate the resistance between A and B for the resistor network in figure 4.7.

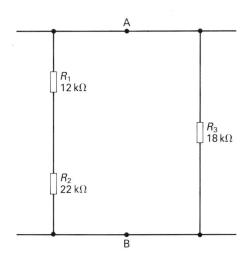

Figure 4.7 Resistor combination for question 3

4 You are provided with three resistors: A = 1.2 kΩ, B = 2.2 kΩ, C = 5.6 kΩ.

How would you obtain resistances of about:

(a) 6.8 kΩ,
(b) 8 kΩ,
(c) 800 Ω,

from a combination of any two of A, B, and C?

△

▽ 4.6 **Cells in series and parallel**

Having dealt with series- and parallel-connected resistors, it is easier to see the effect of connecting cells and batteries in series and parallel.

Figure 4.8 shows the symbol for a 9 V battery made up of six cells each of e.m.f. 1.5 V connected internally in series with each other. Note that the negative terminal of one cell is connected to the positive terminal of the next; in this way the individual cell e.m.f.s add up to 9 V. Note that not all cells have an e.m.f. of 1.5 V as has the carbon–zinc dry cell.

Suppose you connected two 4.5 V batteries together as shown in figure 4.9. Would any current flow through the resistor? This is a series connection of two batteries, but no current flows since the e.m.f.s are equal and oppose each other. However, if one battery had an e.m.f. of 3 V, as shown in figure 4.10, the battery of higher e.m.f. would force current to flow through the 3 V battery in the wrong direction.

Figure 4.11 shows two batteries connected in parallel. Current flows through the resistor as you might expect, but what is the voltage across R? Not 9 V, but 4.5 V, as you can prove for yourself by using two 4.5 V batteries and a voltmeter.

However, if one of the batteries in figure 4.11 had an e.m.f. of 3 V, what would be the voltage across R? Not 3 V, but the higher voltage 4.5 V, as you could easily show with the two batteries of differing e.m.f.s with R equal to about 100 Ω.

The usefulness of connecting two identical batteries in parallel is that they provide twice the current capacity of either battery.

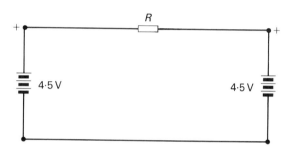

Figure 4.9 Opposing and equal e.m.f.s – no current

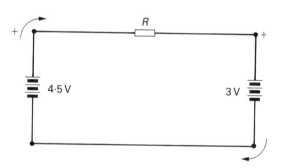

Figure 4.10 Opposing and unequal e.m.f.s – current flows

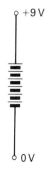

Figure 4.8 Six 1.5 V cells in series provide a 9 V battery

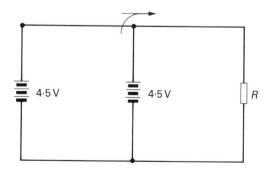

Figure 4.11 Two batteries connected in parallel

▽ 4·7 **Internal resistance**

The *internal resistance* of a battery reduces the current which can be obtained from it, as well as lowering the voltage at the terminals of the battery.

Question

1   Measure the voltage across the terminals of a 4.5 V (or other) dry battery with and without a 10 ohm resistor connected across the terminals. Which gives the higher reading?

The reading without the 10 ohm resistor gives the e.m.f. of the battery. The battery is said to be on 'open-circuit', since it is not delivering current — apart from the very small current required by the voltmeter. Once the battery drives current through the resistor, the voltage drops. The explanation is that the battery has an internal resistance through which current must flow, and, as

you know, when current I flows through a resistor R, the voltage across it is $I \times R$. Thus the e.m.f., less the internal voltage drop, gives the actual terminal voltage. As a battery ages, its internal resistance increases, and it is less capable of providing a useful current and voltage.

The advantage of wet batteries, e.g. lead-acid accumulators, is that they have a very low internal resistance, and large currents can be drawn from them without reducing the voltage across the terminals.

Questions

2   Why are dry batteries not used in cars?
3   A 9 V dry battery gives a terminal voltage of 9 V when not delivering current. When it provides a current of 100 mA, this voltage falls to 8.2 V. What is the internal resistance of the
△   battery?

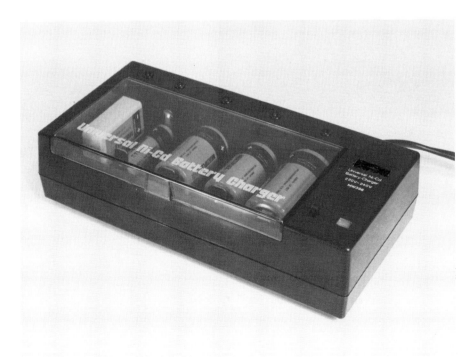

Figure 4.12 A universal nickle-cadmium battery charger — this is the way to save money in the long run
Courtesy: Maplin Electronics plc

5 The Potential Divider and the Wheatstone Bridge

5.1 The potential divider

Figure 5.1 shows how a three-terminal potentiometer (a 3-terminal variable resistor) is used to provide a variable voltage between zero volts and the e.m.f. of the battery. To see how it works, look at the two resistors connected in series in figure 5.2.

Since the same current, I, flows through each resistor, we can see that

$$\frac{V_1}{R_1} = \frac{V_2}{R_2} \text{ and } E = V_1 + V_2$$

Therefore $V_1 = E - V_2$

and $\qquad V_2 = \dfrac{R_2 \times V_1}{R_1}$

or $\qquad V_2 = \dfrac{R_2 \times (E - V_2)}{R_1}$

or $\qquad V_2 = \dfrac{R_2 \times E}{R_1 + R_2}$

Here V_2, the potential difference across R_2, depends on the value of R_2. If the wiper of the potentiometer is moved to the end, A, of the track, R_2 is zero so $V_2 = 0$ V. If the wiper is moved to the end, B, of the track, R_1 is zero and $V_2 = E$, the e.m.f. of the battery. And any value in this range, 0 volts to E volts, can be obtained by moving the wiper along the track.

Thus a *potential divider* can be produced using a potentiometer or two fixed-value resistors. Note that the potential divider can be regarded as a useful building block which takes an input voltage and delivers an output voltage. Since the output voltage is never more than the input voltage, the potential divider is an example of an *attenuator*, i.e. it attenuates or decreases a voltage applied to it.

Let's take a practical example of a potential divider. In figure 5.2, the p.d., V_2, across R_2 is given by

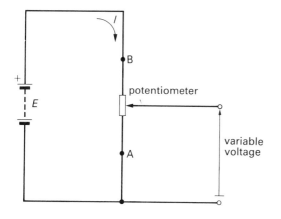

Figure 5.1 Using a potentiometer to provide a variable voltage

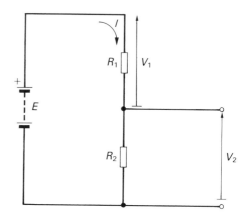

Figure 5.2 Two resistors in series as a potential divider

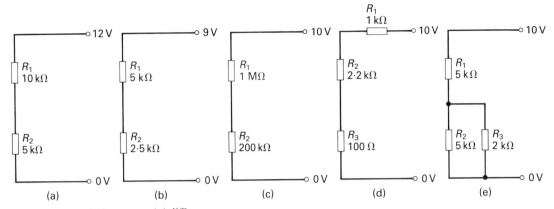

Figure 5.3 Calculating potential differences

Simply by changing the value of R_2 in figure 5.2 it is possible to obtain any voltage between zero ($R_2 = 0$) and the e.m.f. of the battery. In Chapters 6 and 7 it is shown how resistor R_2 can be replaced by a light dependent resistor or a thermistor. In this way it is possible to obtain a voltage which varies with temperature or light intensity, which enables light-operated switches and thermostats to be designed.

Questions

1   Calculate the potential difference across resistor R_2 in each of the circuits in figure 5.3.

2   You have a 100 ohm potentiometer, a 2 V supply and a range of resistors. How can you construct a voltage supply which is variable over the range 0 V to 0.01 V?

▽ 5·2   **Loading a potential divider**

In electronic circuits, the potential divider provides a voltage to make something happen, e.g. it activates a transistor in a light-operated switch. The transistor, or other device, which is operated by the potential divider generally requires a current to make it work. This current has to come from the potential divider. If the

potential divider is to work properly, the current drawn from it has to be as small as possible.

Figure 5.4 shows a potential divider in which the *output p.d.* across resistor R_2 is used to operate a lamp (the *load*). Since the lamp requires current to operate it, the lamp is said to *load* the potential divider since it draws current from it. Note that the p.d. applied across the two resistors in series is called the *input p.d.*

Clearly, the current which lights the lamp has to flow through R_1. So more current flows through R_1 than when the lamp is disconnected. This extra current flow increases the potential difference across R_1 so that the voltage at point X is lowered. If the resistance of the load gets smaller, it tends to draw more current and

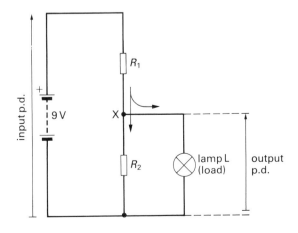

Figure 5.4 The effect of loading a potential divider

this lowers the voltage at point X further. The higher the resistance of the load, the less the voltage at point X is reduced. So if you design a potential divider to give a particular voltage, it is wise not to put a low resistance load on it, otherwise you will get less voltage than you intended.

Remember

The effect of loading a potential divider is to *decrease* the output p.d. from it.

Ideally, the load should not draw any current from the potential divider but in practice this is rarely achieved, so a working rule is generally agreed upon. The rule is that the current drawn by the load should be no more than 10% of the current flowing through the potential divider.

The word *regulation* is used to describe the ability of a potential divider to maintain its output voltage constant under varying load conditions. Regulation is also used to describe the ability of a power supply to maintain a steady voltage as the current drawn from the power supply varies with load current. Regulation is defined as follows:

$$\text{regulation} = \left[\frac{(\text{no-load p.d.}) - (\text{full-load p.d.})}{\text{full-load p.d.}} \right] \times 100\%$$

If the output p.d. is steady, this percentage is small and the regulation is good. There is poor regulation if the percentage is high.

Questions

1   Referring to figure 5.4, if $R_1 = 1$ kΩ, $R_2 = 100$ Ω and R (load) $= 100$ Ω, what is the change in the output p.d. when the load is connected?
2   In question 1, what is the regulation of this potential divider?
3   The p.d. across the output terminals of a potential divider is 50 V when it is not delivering a current. The output p.d.

falls to 48 V when it is delivering a current. What is the regulation provided by this potential divider?

5.3   The Wheatstone bridge

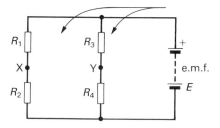

Figure 5.5 A pair of potential dividers connected in parallel

Look at figure 5.5. It shows two potential dividers connected in parallel and operated from the same source of e.m.f., E. The voltage at X is given by

$$\frac{E \times R_1}{R_1 + R_2} = \frac{E}{1 + R_2/R_1}$$

and the voltage at Y is given by

$$\frac{E \times R_4}{R_3 + R_4} = \frac{E}{1 + R_3/R_4}$$

Thus if the voltage at Y equals the voltage at Y, the ratio R_2/R_1 equals the ratio R_3/R_4 and there is no potential difference between the points X and Y. If an ammeter were placed between the points X and Y, there would be no current flow through it.

This combination of two potential dividers supplied from the same source of e.m.f. is known as a *Wheatstone bridge*. This bridge is said to be *balanced* when the above equation holds true. Note that the equation is true for the ratio of two pairs of resistor values. Thus, both R_1 and R_2 might have values higher or lower than R_3 and R_4. But if the ratio of these two pairs of values is the same, the bridge will balance.

There are a number of circuit designs in *Basic Electronics* which use the Wheatstone bridge for measurement and control — see the experiments in Sections 7.5 and 7.9, for example. In these applications of the Wheatstone bridge, one of the resistors, R_1 say, has a resistance which varies with temperature, or light, or pressure, for example. Such a device is called a *transducer* and the light dependent resistor and thermistor are examples of transducers — see Chapter 7 for an introduction to these devices. The Wheatstone bridge is first balanced. Then, if any change occurs in the resistance of the transducer, a voltmeter or electronic amplifier senses this 'out-of-balance' voltage and gives an indication of the change in the quantity (e.g. temperature).

Note that the above equation can be written as

$$R_1 = R_2 \times R_3/R_4$$

Questions

1 Refer to figure 5.5. Suppose the bridge is balanced and that R_1 is a resistor of unknown value. If $R_2 = 120\ \Omega$, $R_3 = 1\ k\Omega$ and $R_4 = 220\ \Omega$, what is the value of R_1?

2 Referring to question 1, suppose the value of R_1 increases slightly after the Wheatstone bridge has been balanced. What happens to the voltage at point X? Does it rise above or fall below that at point Y?

6 Electrical Energy and Power

6.1 Introduction

There are two sources of electrical energy for operating electronic circuits. One of these sources is the 240 V *mains supply* of alternating current which can be converted into a d.c. voltage (e.g. 9 V) suitable for operating most of the circuits in *Basic Electronics*. The other source is *dry batteries* which need replacing every now and then as they become exhausted. Some batteries are *rechargeable* so that they need not be thrown away until they have been used many times.

Electrical energy is used by circuits in a number of ways. For example, an amplifier increases the power of electrical signals so that a loudspeaker can be operated. This is the conversion of electrical energy into useful mechanical energy.

However, in electronic circuits, the conversion of electricity into heat isn't wanted. Heat represents wasted energy so it is necessary to keep to a minimum the heat produced in components and circuits. Integrated circuits, in particular, can easily be damaged by excess heat produced in them, since they contain a very large number of electronic components packed together on a silicon chip. A component which produces heat is said to *dissipate* heat. In this chapter you will see how it is possible to calculate the amount of heat dissipated in a component through which current flows.

6.2 The joule

All forms of energy are measured in a unit called the *joule*. The following examples will enable you to get a rough idea of its size.

(a) The energy required to lift a 1 kg book from the floor onto a table 1 metre above it is about 10 joules (figure 6.1).

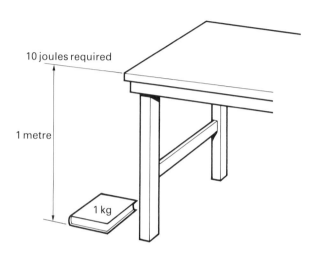

Figure 6.1 Energy required to lift a book

(b) The electrical energy being converted into heat and light by a 100 watt bulb every second is about 100 joules (figure 6.2).

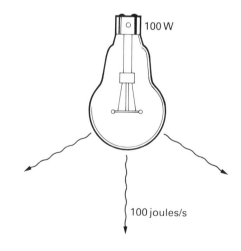

Figure 6.2 Energy emitted from a bulb

(c)   The electrical energy converted into heat in boiling a standard kettle of water is about half a million joules (figure 6.3).

500 000 joules
to boil kettle

Figure 6.3 About half a million joules are required to boil a kettle

6.3   **The watt**

Whereas energy is measured in joules, power is measured in *watts*. You will be familiar with the watt, since it is used to rate electrical devices, for example, a 100 W lamp, a kilowatt (1 kW) fire, or a 7.5 kW oven.

Power measures the rate at which energy is being converted into another form. A 100 watt lamp converts electrical energy into heat and light at the rate of 100 joules per second; a 750 watt electric motor converts electrical energy into mechanical energy (and some heat and sound energy) at 750 joules per second; and so on.

Power (watts) = energy (joules) per second
1 watt = 1 joule per second
1W = 1 J/s

Using this relationship between watts and joules enables you to calculate the energy converted by an electrical device in any known time. For instance, a 100 W lamp run for 10 minutes (600 seconds) uses up 100 × 600 or 60 000 joules of electrical energy.

Suppose a 100 W lamp is used for 10 hours. The electrical energy used is 100 × 10 × 3600 (since there are 3600 seconds in 1 hour) = 3.6 million joules. This large quantity of energy is known as a *kilowatt hour*, since it corresponds to a power of 1 kW operating for 1 hour. Electrical energy is paid for by the kilowatt hour, known more familiarly as the electrical *unit*.

You can easily find how many units of electrical energy are used by a 2 kW (2000 W) electric fire left switched on for 4 hours. Since 1 kW for 1 hour is 1 unit, 2 kW for 4 hours is 8 units. Therefore:

kWh (or units) = watts × hours/1000

Questions

1   A soldering iron (figure 6.4) is rated at 20 W. How many joules of heat energy does it produce in 15 seconds?

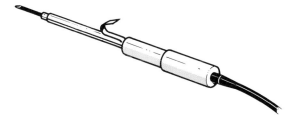

Figure 6.4 A low-wattage mains soldering iron

2   A mains-operated microcomputer is rated at 50 watts. How many hours would you operate it for, to use up 1 kWh?

▽6.4   **Calculating electrical power**

A resistor converts electrical power into heat but how much heat energy is produced in a resistor per second? This equation allows you to calculate that energy:

watts (joules/second) = amperes × volts

or $P = I \times V$

Thus, if $V = 9$ V and $I = 0.5$ A in figure 6.5 overleaf, the electrical power dissipated as heat in the resistor is given by 0.5 × 9 = 4.5 W or 4.5 joules/second.

Figure 6.5 $P = V \times I$

Knowing V and I enables the resistance of R to be found: $R = V/I = 9/0.5 = 18$ ohms. And using this equation enables us to write $P = I \times V$ in two other ways.

Question

1 Can you show that:

$$P = I^2 \times R$$

and $$P = V^2/R?$$

Example 1

What is the electrical power dissipated in a resistor of 1 kΩ when a current of 20 mA flows through it?

$$W = I^2 R = \left(\frac{20}{1000}\right)^2 R = \left(\frac{20^2}{1000^2}\right) \times 1000 = \frac{20^2}{1000}$$

$$= 400 \text{ mW (400 milliwatts)}$$

Example 2

A wire-wound 100 ohm resistor is rated at 3 W. Is it safe to connect it across a 20 V battery?

$$W = \frac{V^2}{R}$$

$$\therefore W = \frac{20^2}{100} = 4 \text{ W}$$

Thus the answer is 'no'. This is 1 W higher than the permitted maximum, and there is a possibility of the resistor burning out.

Questions

2 An electric-fire bar passes a current of 4 A when connected to the 250 V mains. What is the power rating of the bar?

3 A car rear-light bulb carries a current of 1.75 A when connected to a 12 V battery. What is the wattage of the bulb?

4 A soldering iron has a power rating of 16 W. What is the current flowing through the element when it is connected to 240 V mains? What is the resistance of the element?

5 A 9 V dry battery has an internal resistance of 30 ohms and is delivering a current of 60 mA. What is the electrical energy dissipated as heat *inside* the battery? What is the voltage across the terminals of the battery when it is delivering this current?

6 What are the origins of the words *joule* and *watt*?

6.5 **Power ratings of resistors**

The maximum electrical power to be dissipated in electronic components is always specified by the manufacturer. It is essential to ensure that a component for a circuit is operating within its permitted maximum wattage.

Resistors are rated at values of $\frac{1}{20}, \frac{1}{8}, \frac{1}{4}, \frac{1}{2}$, 1 W and upwards. If you look at an illustrated components catalogue, you will see that high-wattage resistors are generally larger than low-wattage types. Fixed-value and variable wire-wound resistors are used generally for high-power (greater than 1 W) applications, whereas moulded-carbon, metal-film, etc. are used for low-power applications (less than 1 W).

Questions

1   A 10 kΩ resistor carries a current of 1 mA. What is the voltage across it and the power dissipated by it?

2   Is a $\frac{1}{4}$ W, 1.2 kΩ resistor suitable for a circuit in which the voltage across it is 10 V?

3   If you find a 4.7 kΩ resistor becoming too warm in a circuit because its power rating is not high enough, would it help to replace it by:

(a)   two 2.2 kΩ resistors in series? or
(b)   two 10 kΩ resistors in parallel?

Each of these resistors has the same power rating as the 4.7 kΩ resistor.

6.6   Using a light emitting diode

The light emitting diode (LED) is used as an indicator lamp in many circuits in this book. A common type of LED is shown in figure 6.6 together with its symbol. When current passes through the LED from its anode terminal to its cathode terminal, it emits light. You can buy LEDs which emit red, green, yellow or blue light. A red LED

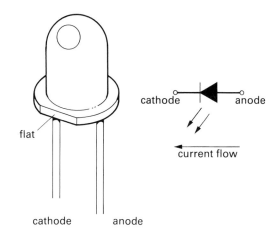

Figure 6.6 A typical LED and circuit symbol

is often used on electronic equipment to indicate 'power on' or 'danger'.

It is important to make sure that too much current does not flow through an LED otherwise it will burn out. An LED will happily work from a 1.5 V battery without damage, but a 9 V battery will destroy it. For power supply voltages higher than 1.5 V, a resistor must be connected in series with the LED to limit the current flowing through it as in figure 6.7. The value of the resistor required is calculated using the equation $R = V/I$.

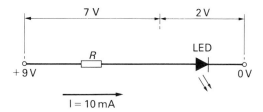

Figure 6.7 Calculating the value of the series resistor

Suppose the LED is to be operated from a 9 V supply (figure 6.7), and the data sheet for the LED states that the current flow through the LED should not be above 20 mA. Let's limit the current to 10 mA to be on the safe side — it will still give a good light output with this smaller current. The manufacturer's data sheet also states that the potential difference across the diode when this current flows through it is 2 V. What should be the value of the resistor, R, in series with the LED?

The potential difference across the series resistor must be

(9−2)   V = 7 V

The current through the resistor is the same as the current through the LED. Therefore the value of the resistor is given by

$R = (V/I) = (7 \text{ V}/10 \text{ mA}) = 700$ ohms

Now it doesn't matter that you have exactly this value so a 680 ohm resistor would be suitable.

The power rating of the resistor can now be calculated from the equation

$$P = V \times I = 7 \text{ V} \times 10 \text{ mA} = 70 \text{ mW}$$

so a ¼ watt resistor is quite suitable.

Questions

1 What would be the value of the resistor required if the LED is to be operated from a 15 V supply? (see figure 6.7)

2 Suppose you were to operate two of the LEDs in series with each other from a 9 V supply. What would be the value of the resistor?

Since current cannot flow through the diode from cathode to anode, no harm will be done to it if it is connected the wrong way round in a circuit (i.e. cathode to the positive supply). Many LEDs have their cathode terminal marked by a 'flat' (see figure 6.6) on the rim round its edge, or one terminal is longer than the other. See Chapter 11, Book C, for more details about the use of the light emitting diode.

▽ 6.7 **Proving that power = volts × amps**

Suppose that there is quantity of easily moved charge, Q coulombs, in the length, l metres, of wire shown in figure 6.8. This

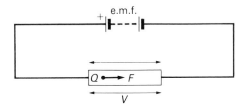

Figure 6.8 Force F on a charge Q in a conductor

charge moves, and a current therefore flows, when there is potential difference of V volts across the ends of the wire. Remember that a current of 1 ampere is a rate of flow of charge of 1 coulomb per second.

Clearly, to make the charge move it must be acted on by a force. This force comes from an electric field which is produced in the wire by the electric potential difference across the ends of the wire. This behaviour is similar to the gravitational potential difference which must exist between the top and bottom of a hill if an object is to move down the hill under the action of a gravitational force.

The value, N, of the electric field which makes the charge Q coulombs move is defined as

$$N = \frac{\text{electric potential difference}}{\text{length of wire}}$$

$$N = \frac{V}{l}$$

and is therefore measured in volts per metre (V/m).

Now the force, F newtons, which acts on the charge Q coulombs, is defined as

$$F = \text{electric field} \times \text{electric charge}$$

$$F = N \times Q$$

$$F = \frac{V \times Q}{l} \text{ newtons}$$

And the work done, W joules, by the electric field in moving this charge from one end of the wire to the other is given by

$$W = \text{force} \times \text{length of wire (i.e. newtons} \times \text{distance)}$$

$$W = F \times l$$

$$W = \frac{V \times Q \times l}{l}$$

i.e. work done = volts × coulombs.
Now we can prove the equation

$$\text{power} = \text{volts} \times \text{amps}$$

Since current = coulombs per second, i.e.

$$I = \frac{Q}{t}$$

we can write $Q = I \times t$. The equation for the work done by the electric field now becomes

$W = V \times I \times t$ joules.

and the rate at which work is done, i.e. the power, P watts, is given by

$$P = \frac{\text{work done}}{\text{time}} \text{ joules per second}$$

$$P = \frac{V \times I \times t}{t}$$

$P = V \times I$ watts.

This is the equation that was used in Section 6.4 for calculating the heat energy dissipated in a resistor by the current passing through it. Whilst this energy is usually wasted as heat (except in the electric fire), some devices are designed to make use of the electrical energy by converting it into some useful form: for example, the electric motor converts electricity into mechanical energy; the light-emitting diode converts electrical energy into light, and so on.

Questions

1 A current of 1 A flowing in a circuit is reduced smoothly to zero in 50 seconds. What total electrical charge in coulombs flowed through the circuit?

2 What is the electric field strength in a wire of length 1 metre when a battery of e.m.f. 9 V is connected across it?

3 A charge of 0.2 coulombs flows along the wire of question 2 in 10 seconds.
(a)   What is the average current flowing through the wire?
(b)   What is the force on this charge?
(c)   How much electrical energy is converted into heat?

$\triangle$

Figure 6.9 Solar panels provide electric power in remote areas Courtesy: British Telecom

7 Special Types of Resistors

7.1 Introduction

This section introduces three special types of resistor which you cannot avoid using when designing circuits for measurement and control applications. These resistors are known as *resistive transducers* because their resistance changes in response to some property.

The *light dependent resistor*(LDR) has a resistance which changes with light intensity so that a photographic lightmeter could be designed using it. *The thermistor*, has a resistance which changes with temperature so that a thermostat can be designed using it. The *strain gauge* has a resistance which changes with the force acting on it so that it can measure how much objects stretch or bend.

7.2 What a light dependent resistor looks like

A common type of light dependent resistor is shown in figure 7.1 but there are other types available from suppliers of electronic components. A light dependent resistor has a window on its flat face under which lies a grid of material making up the resistor. This has a resistance which changes with the amount of light reaching it. The material commonly used is *cadmium sulphide (CdS)* which is a semiconductor — see Chapter 12 in Book C.

The LDR belongs to a class of light detectors generally known as *photocells*, although the term photocell is best reserved for the solar cell. Note that the LDR does not produce electricity from light like the solar cell does. The symbol for the LDR is shown in figure 7.2.

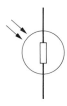

Figure 7.2 The symbol for the light dependent resistor (LDR)

Figure 7.1 A popular LDR — the Mullard ORP12
Courtesy: Mullard Ltd

7.3 *Experiment* B3

Measuring the resistance of an LDR

Set up a multimeter to measure resistance on its 'ohms × 1k' range as explained in Section 3.9. Connect the LDR to the multimeter as shown in figure 7.3(a). Cover and uncover the front of the LDR to change the amount of light reaching it, and note that the resistance of the LDR changes.

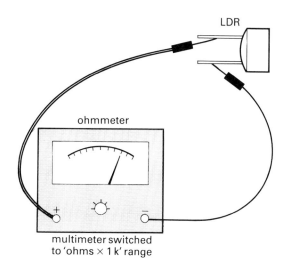

Figure 7.3 (a) Experiment B3: Measuring the resistance of an LDR

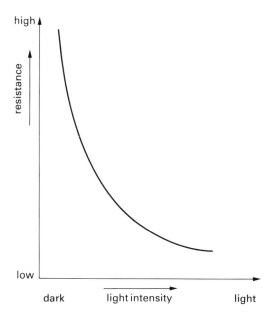

Figure 7.3 (b) Graph showing the variation of the resistance of an LDR

The graph in figure 7.3(b) shows the variation of resistance of an LDR from dark to light conditions. Note that the change is not linear, i.e. the resistance is not proportional to the strength of the light.

Questions

1 What happens to the resistance when the LDR is strongly illuminated?
2 What is the maximum resistance of the LDR?
3 Which part of the LDR is most sensitive to light?

The resistance of an ORP12 is much higher than 1 million ohms (1 MΩ) when it is in the dark, but will fall to less than 1 kΩ when illuminated.

7.4 *Experiment* B4
Using an LDR

In most applications of the LDR, it is necessary to convert the resistance change into a voltage change using a voltage divider.

Use a length of terminal block, or a breadboard, to connect up the simple series circuit shown in figure 7.4. The LDR is

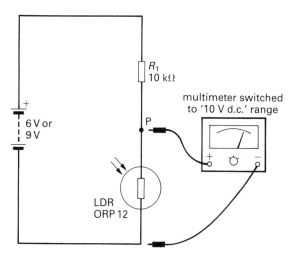

Figure 7.4 Experiment B4: Using an LDR

connected in series with the fixed-value resistor, R_1, so the potential difference across the LDR, i.e. the voltage at point P, varies as the resistance of the LDR varies. Measure this variation using a d.c. voltmeter switched to its 10 V f.s.d. range.

Note that the voltage at point P increases when the LDR is covered, and decreases as light falls on it. This is easily explained if you remember from Section 5.1 that

$$\frac{\text{resistance of LDR}}{\text{resistance of } R_1} = \frac{\text{p.d. across LDR}}{\text{p.d. across } R_1}$$

When the LDR is covered, its resistance is high so the p.d. across it is high; when light falls on it, its resistance is low so that the p.d. across it is low.

This simple voltage divider acts as a *resistance-to-voltage converter* and it is a very useful building block. It is used in Project Module B1 in Chapter 22. The experiment below shows a simple use for two voltage dividers in a simple light meter.

7.5 *Experiment* B5
Making a simple light meter

Use a length of terminal block or a breadboard to assemble the circuit shown in figure 7.5. Note that the circuit is a Wheatstone bridge (Section 5.3) and is made up of two voltage dividers. The milliammeter can be one of the ranges on a

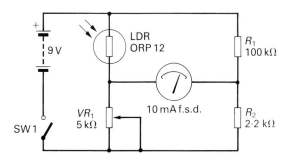

Figure 7.5 Experiment B5: Making a simple light meter

multimeter, any range between 1 mA f.s.d. and 10 mA f.s.d. is suitable.

Fit the LDR in a cardboard tube so that light can reach it only through the window. Cover the end of the tube so that the LDR is in darkness and adjust the variable resistor, VR_1, so that the meter reads zero. Now uncover the LDR and vary the illumination on it. The reading on the milliammeter will vary as the light intensity varies.

The scale of the meter can be calibrated in light units of *lux* by comparing the readings with those of a commercial lightmeter. Alternatively, the calibrations may be in units of photographic f-numbers by comparing its readings with those on a camera lightmeter.

7.6 **What a thermistor looks like**

The thermistor is a resistor made of semiconducting materials having a resistance which decreases as its temperature rises. This type is called a *negative temperature coefficient resistor (n.t.c. resistor)* to distinguish it from the less-common type, known as a *positive temperature coefficient resistor*, whose resistance increases as the temperature rises. Figure 7.6 illustrates the three main types of thermistor: rod, disc, and bead. The symbol for a thermistor is shown in figure 7.7.

7.7 *Experiment* B6
Measuring the resistance of a thermistor

Figure 7.8 shows how to find out how the resistance of a thermistor varies with temperature. A bead thermistor, e.g. type GL16, is suitable for this experiment. The leads of the thermistor should be insulated from contact with the water.

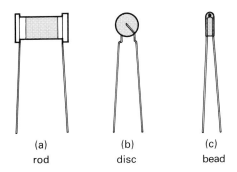

(a)
rod

(b)
disc

(c)
bead

Figure 7.6 A selection of thermistors (a) rod
(b) disc type (c) bead type

Figure 7.7 The symbol for the thermistor

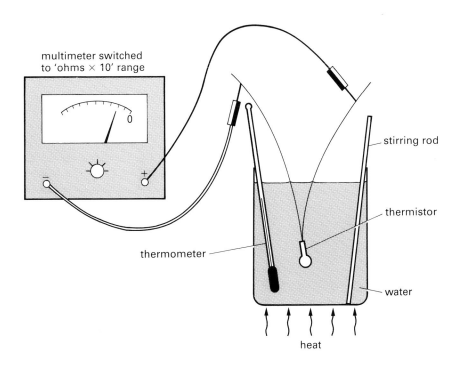

multimeter switched
to 'ohms × 10' range

stirring rod

thermistor

thermometer

water

heat

Figure 7.8 Experiment B6: Measuring the resistance of a thermistor

Use a stirrer, e.g. a glass rod, to keep the water well stirred. A multimeter is used to measure the resistance of the thermistor at different temperatures.

If you plotted a graph of the resistance variation of this thermistor, you would obtain a curve similar to that shown in figure 7.9. Note that the resistance variation with temperature is not linear, i.e. resistance is not proportional to temperature, and this variation is typical only of negative temperature coefficient (n.t.c.) thermistors, i.e. resistance falling with increase in temperature.

Since its resistance varies with temperature, it is usual to identify a thermistor by its resistance at 20°C or 25°C. Thus, the bead thermistors, types GL16 and GL23 have a resistance of 1 MΩ and 2 kΩ, respectively, at 20°C; the rod thermistor type VA1026 has a resistance of 380 Ω at 25°C.

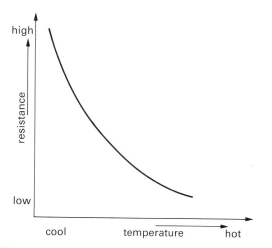

Figure 7.9 Graph showing the variation of resistance of a thermistor

7.8 *Experiment* B7

Using a thermistor in a voltage divider

As with the LDR (Section 7.4), most applications of the thermistor make use of a voltage divider to convert the resistance change into a voltage change.

Repeat Experiment B4 in Section 7.4 using a thermistor in place of the light dependent resistor. The value of R_1 depends on the type of thermistor you use: make sure it has a value about equal to the resistance of the thermistor at 20°C.

The reading of the voltmeter will not change very much as the thermistor is heated and cooled, but electronic circuits can easily be designed to make use of this small change of voltage. This simple circuit, which acts as a resistance-to-voltage converter, is commonly used as part of measurement and control circuits. For example, the next experiment shows a use for two voltage dividers in a simple thermometer.

7.9 *Experiment* B8

Designing a simple thermometer

Use a length of terminal block or a breadboard to connect up the circuit shown in figure 7.10. Note that the circuit is a Wheatstone bridge (Section 5.3) and it is made up of two voltage dividers. The microammeter can be one of the ranges on a multimeter or, if a project is being made up, a stand-alone microammeter. The thermistor should be a glass bead type GL23. As the temperature of the thermistor rises, the voltage at point Y falls and the current through the microammeter increases.

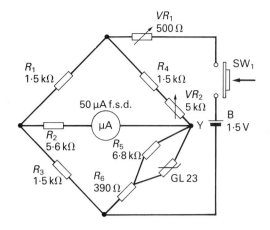

Figure 7.10 Experiment B8: Designing a simple thermometer

moisture reaching its leads. And use a push switch as an on-off switch so that there is no fear of leaving the battery switched on after use. An ordinary pen-cell type 1.5 V battery is suitable for the circuit.

7.10 What a strain gauge looks like

Strictly speaking, there is no such phenomenon as an irresistable force meeting an immovable object. A force, however small, always causes an object to give way slightly: it *distorts* or *strains* in response to the force. For so-called rigid objects, the strain is usually very small so a special type of resistive transducer called a *strain gauge* is used to measure it. A typical strain gauge is shown in figure 7.11 and consists of thin metal foil which has a low resistance — usually between 60 ohms and 2000 ohms.

The thin foil is formed by rolling out an electrically resistive material and etching parts away (rather like a printed circuit board is made) leaving a thin flexible resistor in the form of a grid pattern.

The strain gauge is used by gluing it to the surface of the object undergoing strain. As the object bends, expands or contracts, so does the strain gauge. Now if any metal is stretched, its resistance increases: if it is compressed, its resistance decreases. The resistance change is small, perhaps one tenth of an ohm for a 120 ohm strain gauge — see Section 7.11 — so special electronic

The fixed-value resistors, R_5 and R_6, one in series with and one in parallel with the thermistor, enable the meter to give deflections which are proportional to temperature over the range 0 to 50°C. Variable resistors, VR_1 and VR_2, are adjusted to set the zero point and range of the readings on the microammeter.

To set the zero point, put the thermistor in some crushed and melting ice and adjust VR_2 to make the meter read zero. Place the thermistor in water at 50°C and adjust VR_1 so that the meter reads full scale, i.e. 50°C. Repeat the calibration with the thermistor in melting ice and then water at 50°C and make any further adjustments to the settings if necessary.

If you use this circuit as a working thermometer, make sure that the thermistor is well protected against

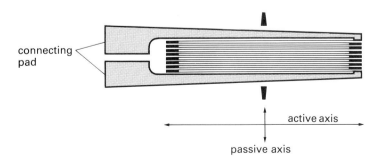

Figure 7.11 One type of strain gauge

circuitry is needed to show the amount of strain on a meter.

This is where the Wheatstone bridge is useful (Section 5.3). As shown in figure 7.12 the strain gauge is placed in one 'arm' of the bridge so that one voltage divider comprises resistors R_1 and R_2, and the other voltage divider comprises resistor R_3 and the strain gauge, R_G. A sensitive voltmeter, V, or electronic amplifier, is placed between X and Y.

If $R_1 = R_2$ and $R_2 = R_G$, the voltages at points X and Y are equal and the voltmeter doesn't register a voltage difference. If the strain gauge is attached to the surface of a material which stretches (called *tensile strain*) under the action of a stretching force (called *tensile stress*), its resistance increases slightly. The voltage at Y increases slightly above that at point X and the voltmeter reads a small value.

In practice, the small change in voltage between points X and Y is amplified by special electronic instrumentation circuits. You will find the design of such a circuit in Chapter 29 of Book C (Project Module C7) which indicates whether the strain experienced by an object is compressive or tensile.

A strain gauge is said to have an *active axis* and a *passive axis* as shown in figure 7.11. As you might expect, the change in strain along the active axis causes a greater change in resistance than the same strain

along the passive axis. In use, the active axis is lined up with the direction in which the strain is to be measured. For example, the strain along the top surface of the beam is being measured in figure 7.13.

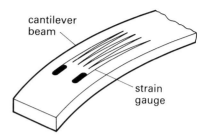

Figure 7.13 The position of a strain gauge on a beam

▽ 7.11   Gauge factor

The gauge factor, G, of a strain gauge is the fractional change of resistance of the gauge, divided by the fractional change in length of the gauge along the active axis:

$$G = \frac{r}{R} \bigg/ \frac{l}{L}$$

r is the change in the gauge resistance R
l is the change in length of the gauge of length L

But l/L is the strain, e, of the object to which the gauge is attached and has no units since it is a ratio of two lengths. Therefore,

$$\frac{r}{R} = eG$$

which shows that the fractional change in resistance of the gauge is equal to the strain along the active axis multiplied by the gauge factor. For most gauges, the gauge factor has a value in the range 1.8 to 2.2.

If we know the gauge factor, G (say 2), the strain, e, that it undergoes (say 5 × 10^{-4}) and the resistance, R, of the strain

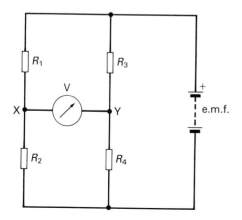

Figure 7.12 A strain gauge in a Wheatstone bridge

gauge (say 120 ohms), we can work out the change in the strain of the gauge as follows:

$$r = R \times e \times G = 120 \times 5 \times 10^{-4} \times 2$$
$$= 0.12 \text{ ohms.}$$

Now it is possible to prove that the change in the potential difference (the output voltage, V) between the points X and Y of the Wheatstone bridge circuit shown in figure 7.12 is equal to

$$V = \frac{E \times r}{4R}$$

to an accuracy of about 1%. Note two points about this equation:

(a) the change in output voltage, V, is proportional to the change in resistance, r, of the strain gauge, and hence to the strain it is responding to and
(b) the output voltage is proportional to the supply voltage, E.

Thus, for small changes of strain, V is proportional to r and hence proportional to strain. Now since

$$\frac{r}{R} = eG$$
$$V = \frac{EeG}{4}$$

that is, the output voltage, V, is
△ proportional to the strain.

▽ 7.12 **Temperature effects**

A change of temperature affects the resistance of electrical conductors (remember the thermistor) and the metal foil of the strain gauge is no exception. An increase in temperature increases the resistance of the gauge so that this adds to the effect of tensile strain. Thus the output voltage from a Wheatstone bridge is higher than it would be if the temperature were constant.

Figure 7.14a shows a technique for compensating for the effects of temperature. A dummy gauge, R_{GI}, is put in place of

resistor R_3 (see Figure 7.12). This gauge is placed close to the gauge R_G which is undergoing strain but it is not itself strained. Its temperature is therefore the same as R_G's temperature. It is possible to show mathematically that a change of temperature does not now affect the output voltage change produced by the Wheatstone bridge. In fact, now the output voltage is

$$V = \frac{E \times r}{2(2R + r)}$$

Because of the problem of ensuring that both the strained and unstrained gauges are at the same temperature, and the bother of having to produce an unstrained specimen of the material, it is usual to mount the dummy gauge on the same specimen as the measuring gauge, but with its active axis at right angles to the direction of strain as shown in Figure
△ 7.14b.

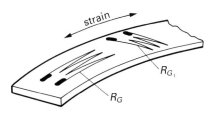

(a) Using a dummy gauge on a separate beam

(b) Dummy gauge on the same beam

Figure 7.14 Compensating for temperature change

7.13   **Why measure strain anyway?**

What is the point of knowing about strain? Take the electricity pylon of Figure 7.15 as an example of a structure which is being strained by the forces acting on it. An engineer will have designed the pylon to withstand all the forces it will experience in its long and lonely life; it has to stand not only its own weight but the effects of wind and ice, and the pulls of the cables it supports. Knowing the elastic properties of steel, the engineer will be able to predict, from an estimation of how large the various forces are on a member of the pylon, exactly what strain is produced (compressive/tensile) in each member. More important, the design should ensure that, even under the most adverse weather conditions, no member can fracture and weaken the pylon. Once a test pylon is built according to the designer's plans, instruments are used to test the type and magnitude of the strains in its members when the pylon is acted upon by various forces. The device which is attached to the pylon is the strain gauge.

Figure 7.15 Electricity pylons Courtesy: National Grid Division, C.E.G.B.

8 What Capacitors Do

8.1 Capacitors store electric charge

A capacitor is put to good use in electronics since it is able to store an electric charge. You should remember that electric charge is measured in coulombs — see Chapter 2.

Questions

1 What is the difference between the coulomb and the ampere?
2 How many coulombs of electric charge move if a current of 300 mA flows for 3 seconds?

8.2 Units of capacitance

A capacitor is said to have a 'capacitance' because of its ability to store charge. Capacitance is measured in *farads*. The farad is a rather large unit, so it is usual to use the *microfarad* (μF or 10^{-6} F, for short), the *nanofarad* (nF or 10^{-9}F), and the *picofarad* (pF or 10^{-12}F). These values are a millionth of a farad, a thousand-millionth of a farad, and a billionth of a farad, respectively.

Suppose a capacitor is marked '470 pF'. You could describe this as a 'four-seventy picofarad' or a 'four-seventy puff' capacitor for short. Alternative ways or writing 470 pF would be 470×10^{-12} F, or 0.00047 μF, or 0.47 nF,

Question

1 Write down the following capacitor values as μF, nF, or pF, whichever seems to be most suitable:

10^{-5} F; 0.0004 F; 10^{-3} F; 5×10^{-9} F; 3.3×10^{-12} F

An alternative way of writing down the values of capacitance is as follows: the multiplier, e.g. micro (μ), nano (n) or pico (p) is written in the place of the decimal point and the letter 'F' is not used. Thus a 470 pF capacitor is written as 470p, a 4.7 μF capacitor is written as 4μ7, a 220 nF capacitor is written as 220n or μ22, and so on.

8.3 The symbol and structure of a capacitor

You will be meeting three symbols for capacitors in circuit diagrams. These are:

the *unpolarised capacitor* (it doesn't matter which way round it is connected in a circuit); its symbol is

the *polarised capacitor* (it has to be connected so that its positive terminal is connected to the more positive potential in a circuit); its symbol is

the variable capacitor which is used in radio tuning circuits and other high frequency circuits; its symbol is

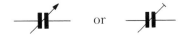

A capacitor, able to store electric charge, is formed if an electrically insulating material, called a *dielectric*, is sandwiched between two metal plates as shown in figure 8.1.

The *capacitance*, C farads, of this capacitor is the amount of electric charge, Q coulombs, it can store per volt of potential difference, V, across the plates, i.e.

capacitance = coulombs/volts
$$C = Q/V$$

The capacitance, C, is actually determined by the physical size of the capacitor. It is directly proportional to the area, A, of the metal plates, and inversely proportional to the separation, d, of the plates. It is also dependent on the type of dielectric between the plates. We can write:

capacitance $C \propto A/d$
($\propto$ means 'proportional to')

The larger A is, and/or the smaller d is, the higher the value of the capacitor. Thus large value capacitors, e.g. electrolytic types, generally have a large size since the area of the plates must be large.

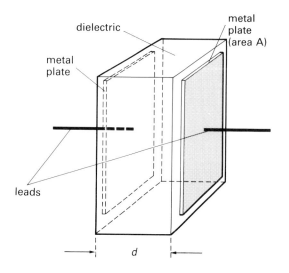

Figure 8.1 Structure of a capacitor

8.4 Operating characteristics of a capacitor

Breakdown voltage

This is the maximum voltage which should be applied across a capacitor. It is usually marked on a capacitor as the *working voltage*, for example '100 V wkg'. If the working voltage is exceeded, the capacitor may be permanently damaged. Always choose a capacitor which has a working voltage above the voltage which you are going to apply across it.

Tolerance

This tells you the range of values that a particular capacitor could have above and below the stated value. For instance, a 1 μF capacitor with a tolerance of ± 20% could have any value between $(1 - 0.2 \times 1)$ μF and $(1 + 0.2 \times 1)$ μF, that is 0.8 and 1 μF.

Questions

1   A mica capacitor with a tolerance of 1% has a value of 100 pF. What are the upper and lower values which it could have?
2   An electrolytic capacitor has a value of 50 μF, and the tolerance lies in the range shown in the table in Section 8.5. What are the upper and lower values of its capacitance?

Stability

The value of a capacitor changes with age, temperature, and other conditions. A capacitor with good stability is one which keeps within the tolerance range as these conditions alter.

Power factor

This measures the fraction of the energy which is lost in the dielectric of the capacitor. This energy appears as heat during the charge and discharge of the capacitor in an a.c. circuit. The power factor is zero for a perfect capacitor and has a maximum value of 1 for a very poor and useless capacitor. It is low for mica, ceramic, and polystyrene capacitors, which are suitable for high-frequency operation.

Leakage current

The dielectric is an insulator, and current should not flow through it. However, a perfect electrical insulator does not exist, and a small leakage current does flow. A good capacitor has a low leakage current.

Figure 8.2 The appearance of a mica capacitor

8.5 Types of capacitor

The table below lists some of the important types of capacitor and the typical range of values for each.

A typical *silver mica* capacitor is shown in figure 8.2. Its good points are its high breakdown voltage, low leakage current, and small tolerance.

Polystyrene capacitors are generally tubular in shape, as shown in figure 8.3. They have a similar capacitance range to, but are cheaper than, mica capacitors.

Ceramic capacitors are made in a variety of shapes. A disc type is shown in figure 8.4. Their stability is often poor, and some types are very sensitive to heat, but they do have a low leakage current.

The *polyester* capacitor shown in figure 8.5, overleaf, is popular with circuit designers. Their small size and rectangular shape makes them easy to assemble on

Figure 8.3 The appearance of a polystyrene capacitor

Figure 8.4 The appearance of a ceramic disc capacitor

type	capacitance range		breakdown voltage	tolerance
Silver mica	1 pF	to 10 000 pF	500 V	± 1%
Polystyrene	10 pF	to 22 000 pF	150 V	± 5%
Ceramic	10 pF	to 47 F	30 V to 500 V	± 20%
Polyester	10 nF	to 20 μF	250 V	± 20%
Polycarbonate	10 nF	to 10 μF	50 V to 600 V	± 20%
Tantalum	100 nF	to 100 μF	3 V to 100 V	± 20%
Electrolytic	1 μF	to 150 000 μF	6.3 V to 600 V	− 25% to + 100%

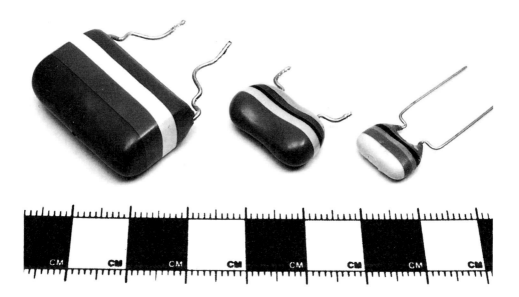

Figure 8.5 Mullard C280 AE series of polyester capacitors Courtesy: Mullard Ltd.

printed circuit boards. These capacitors are often colour-coded as shown in Section 12 in Book A, so it is easy to read their values.

Polycarbonate capacitors have a similar capacitance range to the polyester types, but higher values are available. They are a little more expensive than the polyester types.

Tantalum capacitors are *polarised*. This means that they should be connected in a d.c. circuit as shown in figure 8.6. Note that the symbol for a polarised capacitor is ⎓⊢. Tantalum capacitors have small size, high capacitance, and low leakage current. They are usually made in bead form, as shown in figure 8.7 and their value and working voltage is usually marked on them.

Electrolytic capacitors are mainly tubular or can shaped, as shown in figures 8.8 and 8.9. Their operation depends upon the formation of an oxide film by electrolytic action inside the can as soon as they are

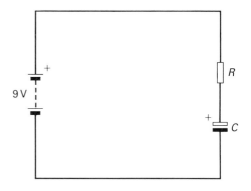

Figure 8.6 How to connect an electrolytic capacitor in a d.c. circuit

Figure 8.7 The appearance of a tantalum bead capacitor

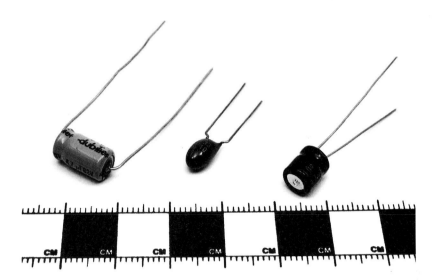

Figure 8.8 Three types of electrolytic capacitors: tubular (axial); tantalum; tubular (radial) Courtesy: Mullard Ltd.

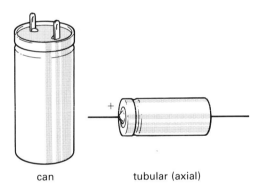

can   tubular (axial)

Figure 8.9 Can and tubular electrolytic capacitors

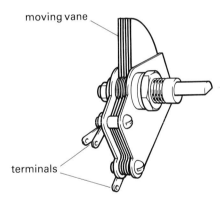

(a) vane type, solid dielectric

connected in a circuit. Note that these capacitors, like the tantalum type, are polarised. If you want high capacitance, then it is the electrolytic (or tantalum) capacitor you look for; but, unfortunately, electrolytic capacitors do have a high leakage current.

Variable capacitors have the symbol

Two useful types are shown in figure 8.10. Variable capacitors are mainly used for tuning in a radio receiver. The capacitance varies with the area of the plates facing

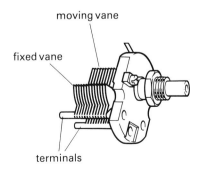

(b) vane type, air dielectric

Figure 8.10 Types of variable capacitor

each other. The moving plates have a special shape so that equal angles of rotation of the spindle cause equal changes of capacitance. Usually the vane type can provide a maximum capacitance in the range 50 to 1000 pF.

Trimmer capacitors are small-value variable capacitors used for fine, preset adjustment of capacitance. Their values do not generally exceed about 150 pF. Trimmers can be the rotating vane type, shown in figure 8.11, or the compression or the concentric type.

The preset capacitance symbol is

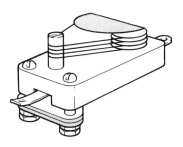

Figure 8.11 An example of a trimmer capacitor

Figure 8.12 shows a particularly interesting type of electrolytic capacitor – a *memory back-up*. This type has a very high value in a small size and is intended to provide power to computer and other equipment in the event of a main power supply failure. The memory back-up shown in figure 8.12 has a value of 3.3 F! It supplies a current of 5.5 μA for 500 hours, or 1 mA for 5 hours at a terminal p.d. of 5.5 V. Whilst in a computer circuit it is continually kept charged by the power supply. Once power failure occurs, the memory back-up discharges providing power for the memory circuits, avoiding loss of data. Book E describes memories in detail.

Figure 8.12 A memory back up capacitor

Questions

1   Open a transistor radio. Look for the tuning capacitor. Is it solid or air dielectric? Can you see any trimmer capacitors? What is their purpose in the radio?

2   Identify any electrolytic capacitors in the radio. How do you know they are electrolytic? What is their working voltage? Are there any other types of capacitor in the radio?

9 Basic Experiments with Capacitors

9.1 *Experiment* B9

Testing a capacitor with a multimeter

Switch a multimeter to the 'ohms × 100' range and connect the black lead to the positive terminal of a large-value electrolytic capacitor — 2000 μF will do. Watch the meter closely while you connect the red lead from the multimeter to the capacitor's negative terminal — see figure 9.1. You will notice the needle swing sharply over to the right and then more slowly move back to the left. Finally the needle will settle down reading a high, steady resistance.

What has happened is this: the internal battery of the multimeter *charges* the capacitor, the rush of charge showing on the meter as a low resistance. The movement of charge is registered as an electric current which eventually falls to zero.

Now remove the meter leads from the capacitor. Switch the meter to the 10 V d.c.

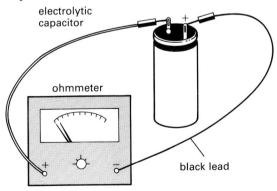

electrolytic capacitor

ohmmeter

black lead

Figure 9.1 Experiment B9: Testing a capacitor with a multimeter

range and, without losing too much time, connect the red lead from the meter to the positive terminal of the capacitor and the black lead to the negative terminal. The capacitor will now discharge through the meter and there will be a steady fall of voltage. This fall may be very slow if the multimeter you are using has a very high resistance.

9.2 *Experiment* B10

Charging a capacitor

The way the voltage changes across a charging capacitor is seen more clearly in this experiment. The simple series circuit shown in figure 9.2 (overleaf) can be wired up on a short length of terminal block or a breadboard. The voltmeter should have a resistance of at least 20 kΩ per volt so use a good quality multimeter or the voltmeter described in Book A (Project Module A4).

Use the wire link to discharge the capacitor. Remove the link and watch the voltmeter carefully. You will see the capacitor charging, since the voltmeter shows a rising voltage across the capacitor. The reading increases quite rapidly at first and then more slowly, finally reaching a steady value. The maximum voltage should be about 9 V if the voltmeter has a high resistance. Electrons are drawn from one plate of the capacitor and pushed onto the other plate by the battery. The build up of electrons on one plate gives rise to the voltage across the plates which is measured by the voltmeter — see figure 9.3 (overleaf).

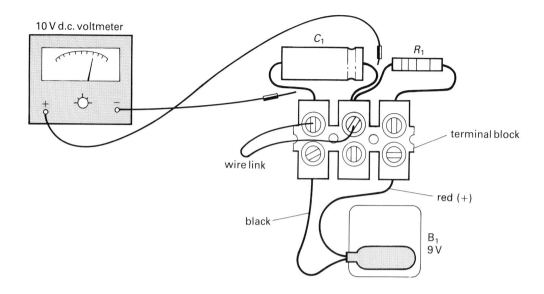

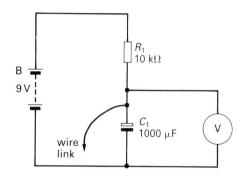

Figure 9.2 Experiment B10: Charging a capacitor

Now quickly take the clip off the battery and use the blade of a screwdriver to short out the two connectors on the clip so that the capacitor discharges through resistor R_1. You will notice that the voltage across the capacitor falls quite rapidly at first, then more slowly, and finally reaches zero. In this case, electrons flow back round the circuit until they neutralise the positive charge on the positive plate. Then there is no potential difference across the plates of the capacitor.

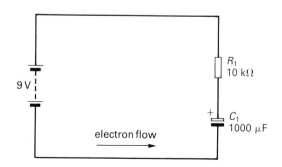

Figure 9.3 Electron flow as a capacitor charges

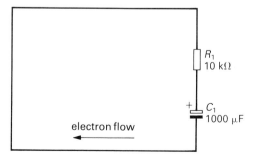

Figure 9.4 Electron flow as a capacitor discharges

9·3 *Experiment* B11

Storing energy in a capacitor

Figure 9.5 shows a simple circuit to show that a capacitor stores energy. The capacitor, C_1, is first charged from the battery, B_1, and then discharged through resistor, R_1, which is in series with a d.c. motor. The value of R_1 should be about 10 ohms. You should use an electrolytic capacitor of at least 1000 μF or connect a number of large-value capacitors in parallel (all positive terminals connected together) so that the total capacitance is equal to the sum of the capacitor values — see Chapter 11.

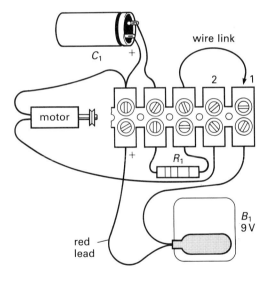

Figure 9.5 Experiment B11: Storing energy in a capacitor

Make sure that the battery voltage is lower than the working voltage of the capacitors you use. Some large-value electrolytic capacitors have a low working voltage, e.g. 6 V.

The wire link is first connected for a few seconds to terminal 1 on the block so that the capacitor is quickly charged by the 9 V battery. Next the wire link is connected to position 2 on the terminal block so that the charge on the capacitor, C_1, flows through resistor, R_1, and through the motor. The motor should run for a few seconds. Experiment with the value of R_1 to obtain the longest running time you can. You could replace the motor by a 6 V, 60 mA lamp so that the lamp glows for a few seconds when the capacitor is discharged through it.

The energy stored even in the largest capacitor is much less than that stored in a single mercury cell, e.g. a hearing aid cell. But large-value electrolytic capacitors are sometimes used in computer circuits to provide a temporary power back-up so that data stored in its memory is not lost in the event of failure of the main power supply (see Section 8.5). Camera flash units use high-voltage capacitors to store the charge which lights the flash-tube. The quantity of energy, in joules, which is stored in a capacitor can be calculated as shown in Section 12.3.

10 Time Constant

10.1 Introduction

The preceding experiments show that a capacitor takes time to charge and discharge through a resistor in series with it. This *CR* combination is a very important building block for it is used in many types of timers and oscillators. The following experiment shows that the rate of charge or discharge of the capacitor depends on the values of both *R* and *C*.

10.2 *Experiment* B12

Measuring time constant

When you have assembled the circuit shown in figure 10.1, make sure that the capacitor is completely discharged by connecting the wire link across the terminals of the capacitor as shown. Now you will need the stop-watch function on a digital watch to measure the reading on the

voltmeter at various times after the wire link has been removed.

Find the time it takes for the voltage across the terminals of the capacitor to reach 1 V after the link has been removed. Use the wire link to discharge the capacitor and repeat the reading of the voltmeter until you know the time as accurately as possible.

Record the time and voltage in the table shown in figure 10.2. Discharge C_1 again,

Figure 10.2 Table for time and voltage measurements

time (seconds)	voltage (volts)
	1
	2
	3
	.
	.
	.

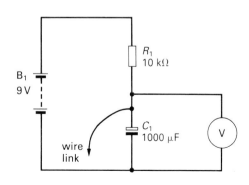

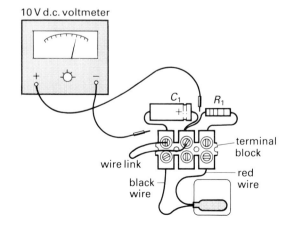

Figure 10.1 Experiment B12: Measuring time constant

and now find the time for the voltage to rise to 2 V, then 3 V, on up to the maximum voltage.

Plot a graph of the voltage across C_1 against time to reach this voltage. Your graph should look like the graph drawn in figure 10.3. Knowing the maximum voltage across the capacitor, you should work out the time for the voltage to reach 2/3 of this value. This is called the *time constant, T*.

For instance, if V_{max} = 9 V, $(\frac{2}{3}) V_{max}$ = 6 V, and T = 10 seconds as you will see from figure 10.3.

The time constant is also given by:

$$T = C_1 \times R_1 \text{ seconds}$$
If C_1 = 1000 μF and R = 10 kΩ, then
$$T = C_1R_1 = 1000 \times 10^{-6} \times 10 \times 10^{-3}$$
$$= 10 \text{ seconds}$$

You should compare this result with the one you have found from your graph.

It is interesting to find the effect on the time constant of changes in the values of C_1 and R_1. You should do this by changing C_1 or R_1 or both. The equation tells you that increasing C_1 or R_1 or both will increase the time constant. Figure 10.4 shows the effect on the time constant of increasing the value of C_1 while keeping R_1 the same value. The smaller the time-constant, the more rapidly a capacitor charges.

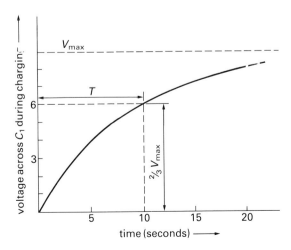

Figure 10.3 Graph showing how the charging voltage varies with time

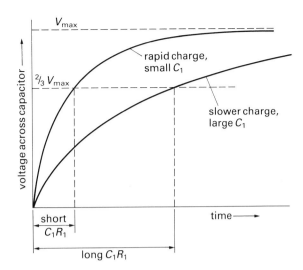

Figure 10.4 Graph showing how the rate at which a capacitor charges depends on the time constant

Question

1   The table in figure 10.5 is incomplete. Can you fill in the missing values?

C_1	R_1	Time-constant C_1R_1
10 μF	–	1 second
0·47 μF	30 kΩ	–
–	1 kΩ	0.2 second
2000 μF	15 kΩ	–
6 8 μF	–	6.8 second

Figure 10.5 Time constant table

10.3 *Experiment* B13

Discharging a capacitor

Suppose you took measurements of voltage against time as the capacitor discharged. You could do this with the circuit of figure 10.1. The graph of voltage against time would be similar to that shown in figure 10.6. The time constant is the time taken

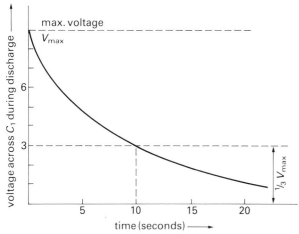

Figure 10.6 Graph showing how voltage varies with time for a discharging capacitor

for the voltage across C_1 to fall by 2/3 of the starting voltage. For $C_1 = 1000\ \mu F$ and $R = 1\ k\Omega$, this is again equal to 10 seconds.

Question

1   Which two pairs of C_1 and R_1 in the following list would give equal time constants if connected in the circuit of figure 10.1.

$C = 2\ \mu F, \quad R = 120\ k\Omega$
$C = 50\ \mu F, \quad R = 10\ k\Omega$
$C = 100\ \mu F, R = 6.8\ k\Omega$
$C = 250\ \mu F, R = 2\ k\Omega$

10.4 The importance of time constant

It should be clear from the way the voltage across a capacitor rises with time (figure 10.3) that the voltage rises more slowly as it approaches the e.m.f. of the battery. In fact theory shows that the p.d. across the capacitor never actually reaches the same value as the e.m.f. of the battery. Therefore time constant is used to find how long it takes the capacitor to charge to a voltage of

2/3 of the e.m.f. of the battery. Given a resistor and a capacitor of known values, you can easily work out the product of these values to get the time constant.

In many circuits, especially oscillators and timers, a resistor is used to charge a capacitor connected in series with it. You will find that the time constant is a useful way of understanding how these circuits work. The astable and the monostable are two circuits which make use of charging capacitors and these are discussed in Chapter 14.

▽10.5 Charge and discharge equations

It is interesting to look at the equations for the charge and discharge of a capacitor. The circuit of figure 10.7 allows you to charge the capacitor, C_1, when the switch SW_1 is in position 1; and to discharge when it is in position 2. The rate at which the capacitor charges and discharges is controlled by resistor R_1.

Figure 10.8 shows both the charge and discharge curves. Notice that for the charging curve, the voltage across C_1 rises from 0 to 6 V in one time constant; after the next time constant, from 6 V to 8 V; and so on. In fact, wherever you choose to start on the curve, the voltage across the capacitor rises to two-thirds of the voltage

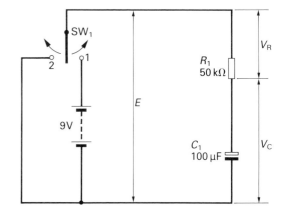

Figure 10.7 Charge and discharge circuit

remaining in a time equal to one time constant. A relationship which has this property is known as an *exponential relationship*.

The time constant for the circuit of figure 10.7 is given by

$$T = C_1 R_1 = 100 \times 10^{-6} \times 50 \times 10^3$$
$$= 5 \text{ seconds}$$

After one time constant the voltage across C_1 from the instant that charging begins, rise to point P, where the voltage across the capacitor is $(2/3) \times 9$ V, or 6 V. After the next time constant, the voltage has risen to $(2/3) \times 3 = 2$ V above 6 V, or to 8 V. After the third time constant, the voltage will have risen to $(2/3) \times 1 = 2/3$ V above 8 V, or $8\frac{2}{3}$ V, and so on. You will be able to see from the falling voltage across the resistor that a similar relation holds for the change of V_R with time.

The equation for the change of voltage with time across the resistor is given by

$$V_R = E \, e^{-t/C_1 R_1}$$

The equation for the change of voltage with time across the capacitor is given by

$$V_C = E \, (1 - e^{-t/C_1 R_1})$$

If your mathematics is good, you will be able to check that these equations do in fact describe what is clear from the curves. For instance, at the moment charging begins, when $t = 0$, these equations give

$$V_R = E \text{ and } V_C = 0$$

When t is very large and the capacitor is fully charged,

$$V_R = 0 \text{ and } V_C = E$$

After the capacitor has been charged and the switch has been moved to position 2, the capacitor will discharge through R_1. Figure 10.9 shows the discharge curves. In this case, note that the current, and hence the voltage, across the resistor is in the opposite direction compared with the charging situation. The lower end of the resistor is now positive compared with the

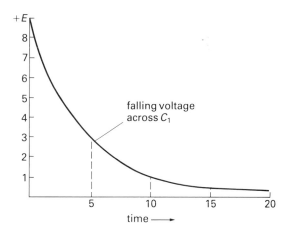

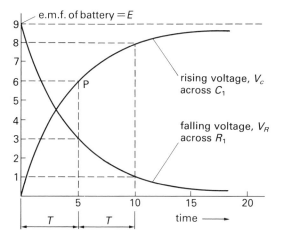

Figure 10.8 Charge and discharge curves for the circuit in figure 10.7

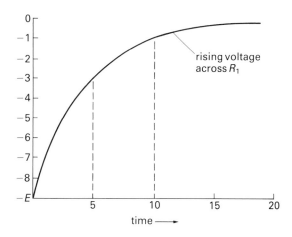

Figure 10.9 How voltage varies across R_1 and C_1 during discharge

top end, and, during discharge, the voltage across C is always in the opposite direction to the voltage across R.

The equations for the voltage changes during discharge are

$V_C = Ee^{-t/C_1R_1}$

and $V_R = E(1 - e^{-t/C_1R_1})$

Once again you can check that these equations describe the graphs by seeing what happens when $t = 0$ and when t is very large. Thus if $t = 0$,

$V_C = Ee^0$

i.e. $V_C = E$ since $e^0 = 1$.

And if $t = \infty$,

$V_C = Ee^{-\infty} = 0$

△ and the capacitor is discharged completely.

▽ 10.6   **Important notes**

The first important point to note is that, during charge, the sum of the voltages across the capacitor and across the resistor

is equal to the e.m.f., E, of the battery. That is,

$E = V_C + V_R$

This should be clear from the graph in figure 10.8.

During discharge the sum of the two voltages is zero, since the battery is not in the circuit. Therefore,

$0 = V_C + V_R$

Or $V_C = - V_R$, which shows that the two voltages are opposing each other.

Finally, note that, when we have talked about the time constant being the time for the voltage to change by 2/3 (0.67) of the starting voltage, we have been using an approximation. By substituting $t = C_1R_1$ in the above equations, you can show that the correct value is nearer to 0.63 than 0.67. Thus if $t = C_1R_1$, the charging equation becomes

$V_C = E (1 - e^{-C_1R_1/C_1R_1})$
$= E (1 - e^{-1}) = E (1 - \frac{1}{e})$

Since e = 2.718,

$V_C = E (1.718/2.718) = E (0.63)$

i.e. V_C is equal to 63% of E after one time
△ constant.

11 Combinations of Capacitors

Introduction

You will sometimes find it necessary to combine capacitors to give a larger or smaller capacitance. Suppose you wanted a capacitance of 500 μF in a circuit, but all you had available were capacitors of value 250 μF. How could you use them to get a value of 500 μF?

From what you have learnt from series and parallel combinations of resistors, you might suggest that you take two 250 μF capacitors and connect them in series. However, this does not give you a value of 500 μF; the two 250 μF capacitors must be connected in parallel in order to give you an equivalent capacitance of 500 μF. This combination is shown in figure 11.1.

terminals) before connecting the black lead of the multimeter to the red (positive) terminal of the capacitor. See how fast the meter needle moves from the right to the left of the scale when the black lead of the meter is connected to the other terminal of the capacitor.

Now discharge the capacitor and connect another 2000 μF capacitor in parallel with the first, as shown. Repeat the observation of the meter needle. You will find that the needle moves more slowly. This must mean that the capacitance of the parallel combination is greater than that of the one capacitor alone — the greater the capacitance, the longer the time that is required to charge it.

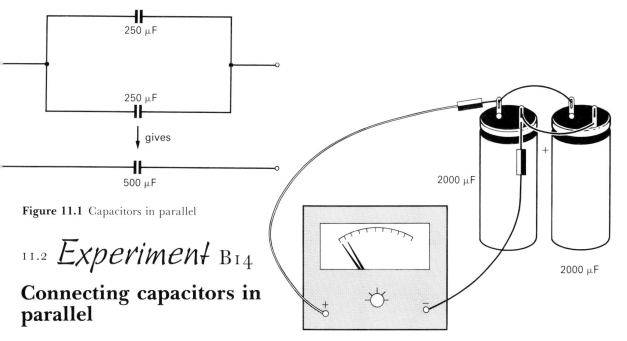

Figure 11.1 Capacitors in parallel

Experiment B14

Connecting capacitors in parallel

You will need the arrangement shown in figure 11.2. Using first just one capacitor, make sure it is discharged (short the

Figure 11.2 Showing the effect of connecting capacitors in parallel

11.3 The formula for capacitors connected in parallel

The rule for calculating the equivalent capacitance of two capacitors connected in parallel is given by the equation

$$C = C_1 + C_2$$

You simply add any additional capacitance connected in parallel, so that for four capacitors

$$C = C_1 + C_2 + C_3 + C_4$$

11.4 *Experiment* B15

Connecting capacitors in series

Now repeat Experiment B14, but this time compare the time for one capacitor to charge with the time for two capacitors to charge when they are connected in series. Make sure that the capacitors are connected so that the positive terminal of one is connected to the negative terminal of the other. You will see that the series combination will charge more rapidly than the single capacitor. This means that the combination has a lower capacitance than the one capacitor alone.

11.5 The formula for capacitors connected in series

The rule for calculating the equivalent capacitance of two capacitors connected in series is given by the equation

$$\frac{1}{C} = \frac{1}{C_1} + \frac{1}{C_2} \quad \text{or} \quad C = \frac{C_1 C_2}{C_1 + C_2}$$

For three capacitors connected in series,

$$\frac{1}{C} = \frac{1}{C_1} + \frac{1}{C_2} + \frac{1}{C_3} \quad \text{or} \quad C = \frac{C_1 C_2 C_3}{C_1 C_2 + C_1 C_3 + C_2 C_3}$$

Let us see how these equations are used.

11.6 Examples

Example 1   Figure 11.3 shows three capacitors connected in parallel. What is the equivalent capacitance across the terminals T?

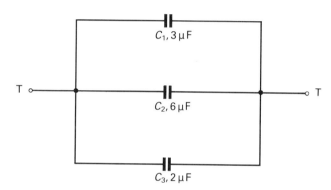

Figure 11.3 Capacitors in parallel

Since the capacitors are connected in parallel, simply add the values together as suggested by the equation in Section 11.3. Thus, the equivalent capacitance equals

$$C_1 + C_2 + C_3 = (3 + 6 + 2)\mu F = 11 \mu F$$

Example 2   Figure 11.4 shows three capacitors connected in series. What is the equivalent capacitance across the terminals T?

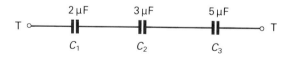

Figure 11.4 Capacitors in series

Since the capacitors are connected in series, you must use the equation given in Section 11.5.

$$\frac{1}{C} = \frac{1}{C_1} + \frac{1}{C_2} + \frac{1}{C_3} = \frac{1}{2} + \frac{1}{3} + \frac{1}{5} = \frac{31}{30}$$

Therefore $C = \dfrac{30}{31}$ or a little less than 1 μF.

Note that, when capacitors are connected in series, the equivalent capacitance is always less than the value of the smallest capacitor.

Example 3   This is going to be a little more difficult. The connections shown in figure 11.5 consist of two capacitors in parallel with another capacitor in series with these two. What is the equivalent capacitance between the terminals T?

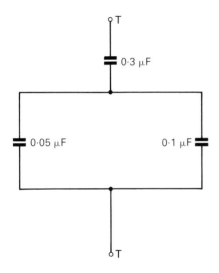

Figure 11.5 Series and parallel capacitors

First, for the two in parallel, the equivalent capacitance is 0.15 μF. This is in series with the 0.3 μF capacitor. The equivalent capacitance is

$$\frac{1}{C} = \frac{1}{0.3} + \frac{1}{0.15} = \frac{3}{0.3}$$

Therefore $C = \dfrac{0.3}{3} = 0.1$ μF

Questions

1   What is the equivalent capacitance of two 2000 μF capacitors connected in parallel?

2   You have one each of the following three capacitors: 2 μF, 5 μF, 10 μF. How would you connect some or all of them to give you equivalent capacitors of values 17 μF; 12 μF; $1\frac{2}{3}$ μF; $3\frac{1}{3}$ μF?

When doing these calculations, you should realise that the values you obtain are not likely to be the values you would obtain when connecting actual capacitors together. This is because of the wide tolerance of some capacitors, particularly the electrolytic type. Thus two 1000 μF capacitors each having a tolerance of 20% could, when connected in parallel, give an equivalent capacitance anywhere between 1600 μF and 2400 μF.

12 Energy Stored in a Capacitor

12.1 Introduction

Since a car tyre expands very little, it is roughly true to say that, if you double the air pressure in the tyre, you double the quantity of air in it (provided the temperature of the air is the same, of course).

Similarly, you can double the quantity of charge stored in a capacitor by doubling the charging voltage — see figure 12.1(a) and (b). You will be able to work out the value of x soon.

It should also be clear to you that a bus tyre has a greater capacity than a car tyre for the same air pressure. So, a 1000 μF capacitor holds far more charge than a 100 nF capacitor for the same voltage applied across the plates.

(a)
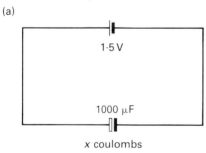
x coulombs

(b)
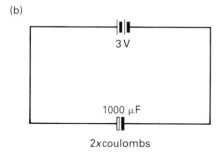
2x coulombs

Figure 12.1 The meaning of charge on a capacitor

Question

1   How many times less charge is stored in the capacitor of figure 12.2(b) than in the capacitor of figure 12.2 (a)?

(a)
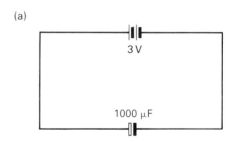

(b)
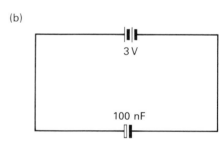

Figure 12.2 Reducing stored charge by reducing capacitor value

You could liken the bursting of a car tyre when the air pressure is too high to the breakdown of a capacitor when the charging voltage is too high.

12.2 The charge stored in a capacitor

It is found that, if you double the charging voltage, the charge stored by a capacitor is doubled. The following equation links the charging voltage with the charge stored and the capacitance of the capacitor:

charge stored = capacitance × voltage
or $Q = C \times V$

One coulomb of charge is stored by a one farad capacitor if it is charged to one volt.

But capacitors are rarely as large as 1 farad (see Section 8.5). For instance, a 1 μF capacitor must store a charge Q for 1 V p.d. across its terminals, where

$$Q = 1 \times 10^{-6} \times 1 = 10^{-6} \text{ coulombs or}$$
1 μC
(one
microcoulomb)

Or the charge stored by a 1000 μF capacitor charged to 10 V is

$$Q = 1000 \times 10^{-6} \times 10 = 10^{-2} \text{ C (one}$$
hundredth
of a
coulomb)

Questions

1 Work out the charge stored in the following examples:
   (a) 4000 μF capacitor charged to 50 V
   (b) 0.05 μF capacitor charged to 6000 V
   (c) 500 pF capacitor charged to 250 V

2 What is the charge x stored by the capacitor in figure 12.1(a)?

12.3 **The energy stored in a capacitor**

Like a tyre under pressure, a charged capacitor is able to do work. If you let air out of a tyre slowly, you could allow the air

to do something useful. Experiment B11 in Chapter 9 showed that a capacitor stores energy, since the electric motor was driven for a short time. It is possible to calculate the energy stored in a capacitor from the following equations:

energy = ½ (charge × voltage)
energy = ½ (capacitance × voltage²)
energy = ½ (charge²/capacitance)

These equations in symbols are

$E = \frac{1}{2}(Q\,V)$
$E = \frac{1}{2}(C\,V^2)$
$E = \frac{1}{2}(Q^2/C)$

Energy is measured in joules. The energy which may be obtained from a 4000 μF capacitor charged to 20 V is given by

$$E = \tfrac{1}{2}(CV^2) = \tfrac{1}{2} \,(4000 \times 10^{-6} \times 20 \times 20)$$
$$= 2 \times 10^3 \times 10^{-6} \times 400$$
$$= 0.8 \text{ joules (J)}$$

which is not a very large quantity of energy. In fact, a current of 1 ampere flowing for one second through a 1 ohm resistor produces 1 joule of heat energy. So even the largest value capacitors store very little energy compared with the energy which may be obtained from a cell.

Questions

1 Calculate the energy which may be obtained from a 2 μF capacitor charged to 400 V.

2 0.001 coulombs of charge are stored in a 100 μF capacitor. What is the voltage across the plates of the capacitor and the energy which could be obtained from it?

13 Capacitors in Alternating Current Circuits

▽ 13.1 **Introduction**

A capacitor does not allow a steady current to flow through it. This is because a capacitor consists of an electrically insulating material (a dielectric) sandwiched between two conducting metal plates to which are attached terminal wires. (See Section 8.3 for the details of how a capacitor is constructed.)

Now a capacitor will charge and discharge if an a.c. supply is connected across it just as though you were charging and discharging it using a battery and a switch. The resulting a.c. current which flows in the external circuit behaves just as if the capacitor were acting as an ordinary resistance to the flow of current. The following two experiments show these effects.

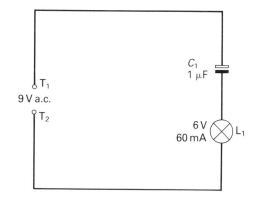

▽ 13.2 *Experiment* B16

Showing that a capacitor passes a.c.

When the simple series circuit shown in figure 13.1 is connected to the low voltage power supply, the lamp lights. It looks as though current flows through the capacitor, but the dielectric of the capacitor is an insulator.

What is actually happening is shown in figure 13.2. Current flows first one way round the circuit and then the other and alternately charges one plate positive and then negative. Current flows through the lamp as though the capacitor were merely

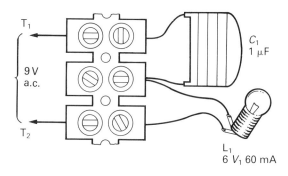

Figure 13.1 Experiment B16: Showing that alternating current flows 'through' a capacitor

an ordinary resistor in the circuit. Note that the capacitor should be a non-polarised type, e.g. a polyester capacitor; use a value of at least 1 μF otherwise the lamp will barely glow.

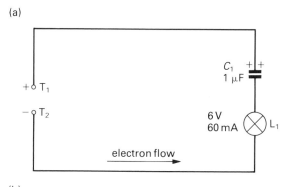

(a)

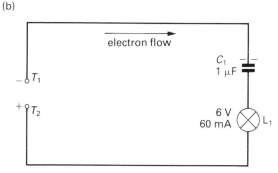

(b)

Figure 13.2 Explanation of how the lamp lights in an a.c. circuit

CAUTION

Do make sure that you use a properly designed low-voltage a.c. supply. Do not tamper with the windings of a low voltage △ mains transformer to get the a.c. supply.

13.3 *Experiment* B17

Measuring the reactance of a capacitor

The simple circuit of figure 13.3 will enable you to take a closer look at the way a capacitor resists the flow of current in an a.c. circuit.

The waveform generator (Chapter 11, Book A) enables the frequency of the a.c. supply to be varied. The a.c. output of the generator should be at least 1 V. The current flowing in the circuit is measured with an a.c. milliammeter, e.g. the 10 mA a.c. range on a multimeter.

Questions

1   Take note of the current flowing 'through' the capacitor for settings of the signal generator of 10, 50, 100, 500, and 1000 Hz. As the frequency increases does the current increase or decrease?

2   If you imagine that the capacitor is an ordinary resistor, how does its resistance change with frequency?

You will notice that the current flowing through the capacitor increases as the frequency increases. The current is lowest at 10 Hz.

3   What would the current be if the a.c. frequency were reduced to zero frequency?

The important conclusion is that the 'resistance' of the capacitor decreases as the frequency rises.

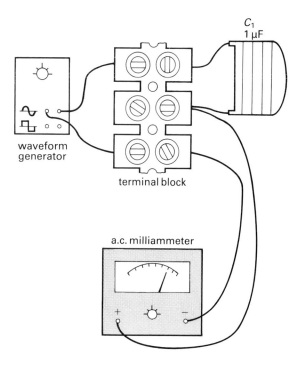

Figure 13.3 Experiment B17: Finding the effective resistance of a capacitor

However, a capacitor is not said to have a 'resistance' but a *reactance*. Reactance is always used to describe the resistance offered by an electronic component to alternating current. In the case of a capacitor, it is known as *capacitive reactance*. This is given by the equation:

capacitive reactance, $X_C = 1/(2\pi f C)$

C is the capacitance in farads, f is in hertz, and X_C is measured in ohms — just as you would measure an ordinary resistance. The equation should bear out what you found from your experiment, that the reactance
△ increases as frequency decreases.

▽ 13.4 Calculating reactance values

It is worthwhile doing a calculation to see how this experiment works. For instance, suppose you wanted to know the reactance of a 100 nF capacitor in the circuit of figure 13.1 when the frequency of the supply voltage is 100 Hz. Then the equation gives

$X_C = 1/(2 \times 3.14 \times 100 \times 10^{-7})$
$= 10^5/6.28 = 16 \text{ k}\Omega$

This is a high reactance because the frequency is low.

Question

1   Calculate the capacitive reactance for the 100 nF capacitor at frequencies of 10, 50, 500, and 1000 Hz.

Use different-value capacitors in the circuit of figure 13.1 and, for a particular frequency — say 500 Hz — prove for yourself that a higher capacitance produces a higher current. This is what you would expect from the equation: increase C and you decrease X_C, and therefore the current increases.

Question

2   In the table shown in figure 13.4 there are some missing values of reactance, capacitance, and frequency. Can you fill in the blank spaces?

	frequency f(Hz)	capacitance C(μf)	reactance X_C(Ω)
(a)	10	100	160
(b)		1	16 k
(c)	10 k	0.1	
(d)	10 k		160 k
(e)		100	1.6
(f)		1	160
(g)		0.1	1.6 k
(h)		1000	16 M
(i)	1 M	1000	0.0016
(j)	1 M		0.16
(k)	1 M	0.1	
(l)	1 M		16 k

△ **Figure 13.4** A table of reactance values for a capacitor

▽ 13.5 Coupling and decoupling capacitors

You have seen in Section 13.2 that a capacitor prevents a direct current from passing through it but it allows an alternating current to flow through it. This property of a capacitor is made good use of in alternating current circuits, for example audio amplifiers. If you look at the Audio Amplifier (Project Module A5) shown in figure 13.5, you will notice that capacitor C_2 allows a.c. voltages from the source of signal, e.g. a microphone or oscillator, to pass to the input (pin 3) of the amplifier, IC_1. But d.c. voltages are prevented from passing through and altering the steady d.c.

voltage which is set on pin 3 of the amplifier by resistors R_1 and R_2. A capacitor used in this way is known as a *blocking* (or *coupling*) capacitor. It blocks the steady d.c. voltage set by R_1 and R_2 but couples the changing a.c. signals from the microphone to the amplifier.

It is important that the coupling capacitor offers low resistance (reactance) to the lowest possible frequency which we want the capacitor to pass. Suppose 50 Hz is the lowest frequency and 100 Ω is the highest resistance we want it to have. Then,

$$C = 1/(6.28 \times 50 \times 100)$$
$$= 32 \text{ } \mu\text{F, approximately}$$
$$\text{(see Section 13.4)}$$

You will find that blocking capacitors in audio frequency amplifiers have values typically 22 μF and higher.

A similar use for a capacitor is to bypass unwanted a.c. signals in circuits. One example of this use is shown in the Radio Receiver (Project Module B5) shown in figure 13.6. Capacitor C_3 bypasses unwanted radio frequency signals, which have carried the music and speech, to 0 V. It has a low value (100 nF) and therefore offers a low resistance to high frequency radio signals but a high resistance to the speech and music which we want to hear. These audio frequency signals are passed to the transistor amplifier based on Tr_2.

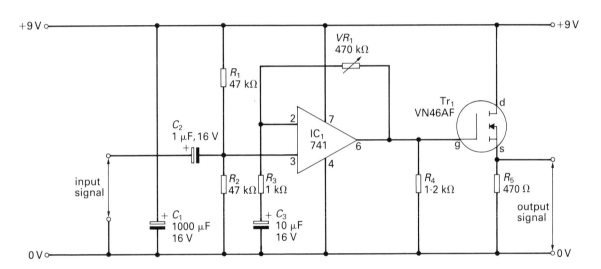

Figure 13.5 The use of a coupling capacitor, C_1 (see Project Module A5, Book A)

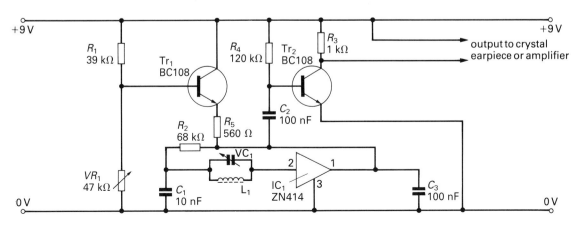

Figure 13.6 The use of a decoupling capacitor, C_3 (see Project Module B5, Book B)

14 The 555 Timer

14.1 Introduction

The *555* (or *triple–5*) is a type of integrated circuit which is designed to work with capacitors and resistors to give time delays. Millions of 555 timers have been made by a number of manufacturers since the first device was produced by an American semiconductor manufacturer, Fairchild Semiconductors, in 1972. Figure 14.1 shows the general appearance of the 555 in the common 8-pin dual-in-line (DIL) package, together with the purpose of its 8 terminal pins. Note that pin 1 is identified as the pin nearest to the small depression or, in some packages, the pin to the immediate left of the notch (viewing the package as shown). A 14–pin DIL package, the '556' containing two 555 timers, is also available for circuits which require two 555s.

There are two forms of the 555 timer: the older *bipolar* type and the more recent *CMOS* type. The table in figure 14.2 compares the specifications of these two devices. Do not try to understand the meaning of all this data at the moment but note two of them. First, the CMOS 555 does not require nearly so much current to operate it as the bipolar 555 which makes it very useful in battery-operated equipment, and second, both devices have a wide supply voltage range.

In this section we shall see two ways of using the 555 timer — as a *monostable* and as an *astable*. Project Module A6 (Astable) and Project Module B4 (Monostable) are based on the 555 timer. Chapter 9 of Book D looks at other uses of the 555 timer and briefly describes the way it works. In this chapter we are interested in how it is used, not how it works.

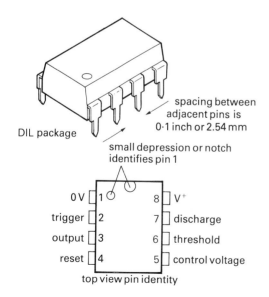

DIL package

spacing between adjacent pins is 0·1 inch or 2.54 mm

small depression or notch identifies pin 1

0 V	1	8	V⁺
trigger	2	7	discharge
output	3	6	threshold
reset	4	5	control voltage

top view pin identity

Figure 14.1 The 555 timer integrated circuit and its pin identity

▽14.2 *Experiment* B18

Designing a monostable using the 555 timer

Figure 14.3 shows the connections to the 555 to make it operate as a monostable circuit. The circuit is easily assembled using breadboard as shown in figure 14.4. The switch, SW_1, is a push-to-make, release-to-break type. Note that the light emitting diode, LED_1, must have a resistor, R_3, connected in series with it. Note that pin 5 is not connected to any part of the circuit.

parameter	CMOS 555	bipolar 555
quiescent current at supply voltage 15 V	typical 120 μA	typical 10 mA
input current for:		
trigger	50 pA	0.5 μA
threshold	50 pA	0.1 μA
reset	100 pA	0.1 mA
maximum operating frequency	500 kHz	500 kHz
power supply voltage range	2 to 18 V	4.5 to 15 V
peak supply current transient	10 mA	370 mA
rise and fall time at output	40 ns	100 ns

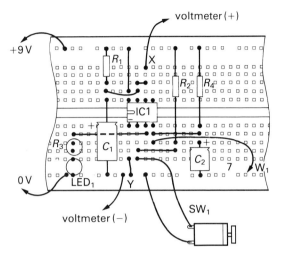

Figure 14.4 Experiment B18: Designing a monostable using the 555 timer

Figure 14.2 The main specifications of the 555 timer

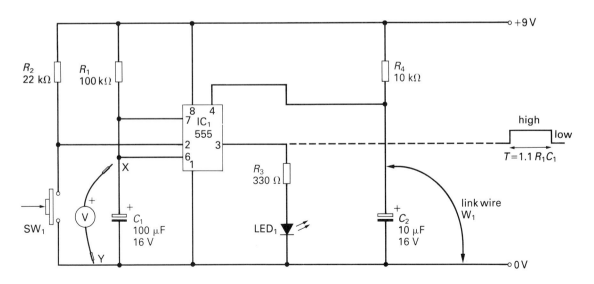

Figure 14.3 The 555 timer wired up as a monostable circuit

When the circuit is assembled and connected to the power supply, make the following observations:

(a)   Press SW$_1$ briefly and release it. LED$_1$, will light and stay alight for a time of about 10 seconds and then go off. Measure this time to the nearest second.

(b)   Double the value of R_1, i.e. connect two 100 kΩ resistors in series, and note the new time for LED$_1$ to remain lit after SW$_1$ is pressed.

(c)   Connect the two 100 kΩ resistors in parallel so that R_1 has a value of 50 kΩ. Note the new time delay after SW$_1$ is pressed.

After the 555 has been 'triggered' by pressing SW$_1$, the time, T, for which LED$_1$ is on is given by the equation

$$T = 1.1 \, R_1 \times C_1 \text{ seconds}$$

This time delay is shown by the rectangular waveform in figure 14.3. Note that the output voltage rises sharply to a voltage near the supply voltage (called the high state of the output) and it falls sharply to near 0 V (the low state of the output voltage).

In the above equation the value of R_1 must be in ohms and the value of C_1 in farads. Since $R_1 = 100$ kΩ $= 10^5$ and $C_1 = 100$ μF $= 10^{-4}$, the value of T is

$$T = 10^5 \times 10^{-4} = 10 \text{ seconds}$$

Now this time is an approximation since we have ignored the factor 1.1 for good reasons. Both the timing components, R_1 and C_1, have nominal values not very close to their real values. This is true especially of the electrolytic capacitor, C_1, which has a wide tolerance so we are justified in using the approximate equation. You will find the time delay is longer than 10 seconds, probably nearer 12 seconds. However, if you repeat the experiment you will find that the delay is repeatable. And, more important, if you change the power supply voltage, for example use a 5 V or 12 V supply, the time delay is unchanged.

When you double the value of R_1 (two 100 kΩ resistors in series) the time delay is doubled. When you halve the value of R_1 (two 100 kΩ resistors in parallel), the time delay is halved as shown by the equation above.

(d)   How would you connect LED$_1$ to the output of the 555 so that it is off during the time delay?

(e)   Pin 4 is the RESET pin. Use the link wire, W, to find the effect of connecting this pin to 0 V during the time delay.

(f)   Change the value of C_1 to 220 μF and explain the effect on the time delay.

(g)   Connect a voltmeter set to 10 V d.c. across the points X and Y in the circuit. Note the rise in voltage across the capacitor after SW$_1$ has been pressed. The voltage rises quickly at first then more slowly until it reaches one-third of the supply voltage, i.e. 6 V if the battery has an e.m.f. of 9 V. At 6 V the 555 suddenly discharges the capacitor and the voltmeter reads zero. The circuit stays in this normally-off state until it is triggered once again by pressing SW$_1$. This normally-off state is the reason why the circuit is called a monostable, i.e. it has one stable state in which the output voltage is zero. The state of the circuit when it is triggered to give a time delay is a temporary △ state.

▽ 14.3 *Experiment* B19

Designing an astable using the 555 timer

Figure 14.5 shows the connections to the 555 to make it operate as an astable circuit. You should assemble the circuit on breadboard, or modify the connections to the monostable circuit shown in figure 14.4. Note that a 47 kΩ resistor is now connected between pins 6 and 7; the switch, SW$_1$, and original resistors, R_1 and R_4, and capacitor, C_2, are not needed, and pin 2 is

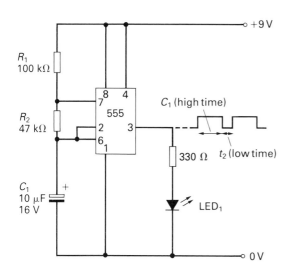

Figure 14.5 The 555 timer wired up as an astable

connected to pin 6. Also the RESET terminal is now connected direct to the positive supply.

When the circuit is assembled and connected to the power supply, you will notice that LED_1 flashes on and off. This is caused by the voltage at pin 3 of the 555 going high and low as shown in the waveform. When the voltage is high, LED_1 is lit; when it is low it is off.

Note that the circuit operates automatically and does not require triggering as with the monostable. 'Astable' means 'not stable' since LED_1 is first on then off, and this cycle repeats as long as the circuit receives power.

Resistors R_1 and R_2, and capacitor C_1 determine how long the output voltage is high and low. The equations for the high time (t_1) and the low time (t_2) are as follows:

high time, $t_1 = 0.7 \times (R_1 + R_2) \times C_1$
low time, $t_2 = 0.7 \times R_2 \times C_1$

Since the resistor, and especially the capacitor, have a wide tolerance, it is usual to ignore the factor 0.7. Thus for the values of components in this circuit:

$$t_1 = (100 + 47 \ )k\Omega) \times 10\mu F$$
$$= 150 \times 10^3 \times 10^{-5}$$
$$= 1.5 \text{ seconds (approx)}$$

and

$$t_2 = 47 \ k\Omega \times 10 \ \mu F$$
$$= 50 \times 10^3 \times 10^{-5}$$
$$= 0.5 \text{ seconds (approx)}$$

(Note that 47 kΩ has been taken as 50 kΩ since we know that the resistor values have a tolerance of perhaps 5%.)

You should be able to see that LED_1 is on about three times longer than it is off, i.e. the astable produces a rectangular waveform with the values of components used in this circuit. The period, T, of the oscillation is the sum of t_1 and t_2. Thus the frequency, f hertz (Hz) of the waveform is given by

$$f = 1/T = 1/(t_1 + t_2) = 1/(2R_2 + R_1) \times C_1$$

In this case, the frequency is

$$\left(\frac{1}{1.5 + 0.5}\right) = \frac{1}{2} = 0.5 \text{ Hz.}$$

(a) How would you use a second LED so that the two flash on and off alternately?

(b) Change the value of C_1 from 10 μF to 100 nF. What effect does this change have on the frequency of the oscillation?

(c) Connect a loudspeaker to the output of the 555 as shown in figure 14.6 so that you can hear the oscillation.

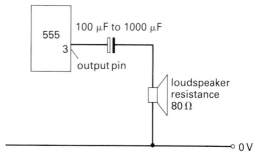

Figure 14.6 How to connect a loudspeaker to the 555 timer

(d) Replace R_1 by a light dependent resistor (LDR). Note how the frequency of the note changes as you vary the illumination of the LDR.

(e) In the circuit shown in figure 14.4 change the value of C_1 to 100 µF and connect a voltmeter, set to 10 V d.c., across the capacitor. Note that the capacitor charges and discharges between about one-third and two-thirds of the supply voltage.

(f) Now replace the 100 µF capacitor by the 100 nF capacitor. Remove the loudspeaker from the output of the 555 and connect an oscilloscope between pin

3 and 0 V. Adjust the oscilloscope to show the rectangular waveform produced by the astable. Measure the frequency of the waveform and compare it with your answer to (a) above.

(g) Also connect the oscilloscope across the capacitor and show that the capacitor charges and discharges between one-third and two-thirds of the supply voltage, i.e. between 3 V and 6 V if you are using a 9 V supply. Is it possible to have the high time (t_2) shorter than the low time (t_1)? If you want some help using the oscilloscope in this experiment, you should read Section 10.7, Book A.

14.4 *Experiment* B20

Designing a timer alarm

The monostable and astable circuits can be used together to produce a system that sounds an alarm after a preset time delay.

The system for doing this is shown in figure 14.7. Building block 1 is the monostable 'programmed' using a resistor and capacitor to give, for example, a one minute delay; e.g. $R_1 = 56$ kΩ, $C_1 = 1000$ µF.

Pin 3 of this monostable is connected

direct to pin 1 of the astable, and pin 3 of the astable provides the signals for the loudspeaker represented by block 3. Use values of $R_1 = 100$ kΩ, $R_2 = 47$ kΩ, and $C_1 = 100$ nF (0.1 µF).

When the system is switched on the alarm sounds. But on pressing SW_1 on the monostable, you will hear the alarm sound after a time delay. How does the system work? Explain why pin 3 of the monostable is connected to pin 1 of the astable.

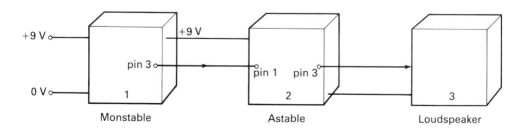

Figure 14.7 The system for the timer alarm

15 What Inductors Do and How They are Made

15.1 What an inductor does

You will remember from the experiments you carried out in earlier chapters that a capacitor blocks the flow of direct current and allows alternating current to flow through it.

Inductors have the opposite effect, for they resist the flow of alternating current but allow direct current to flow. This ability of an inductor to obstruct the passage of alternating current is known as *inductance*. For this reason, inductors are sometimes called chokes because they 'choke' (that is, impede or throttle) the flow of alternating current through them.

Inductance has a part to play in many circuits. For example, radio receivers use an inductor in the form of an aerial coil; transformers use two inductors coupled close together; and metal locators depend for their action on the effect of metal objects changing the inductance of an inductor.

15.2 How an inductor is made

All inductors consist of a coil of wire. The coil may be *air-cored* or wound onto a material which may be easily magnetised and demagnetised, such as *iron* or *ferrite*. These materials greatly increase the inductance of an inductor.

Figure 15.1 shows the structure of an air-cored inductor, the coil being wound on a former made from non-magnetic materials. The electronic symbol for an air-cored inductor is also shown in figure 15.1.

An inductor having an *iron-dust* (or *ferrite*) core is shown in figure 15.2. Here the dust core has been withdrawn completely from the coil. The inductance varies with the length of the core within the coil. The symbol for an iron-dust-cored inductor is also shown in figure 15.2.

Larger-size and larger-value inductors are made in the form shown in figure 15.3, overleaf. Here the coil is wound on a specially shaped core made up of flat iron plates sandwiched together but insulated from each other.

symbol

Figure 15.1 Air-cored inductor and symbol

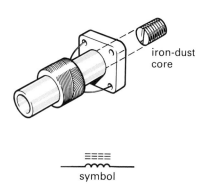

iron-dust core

symbol

Figure 15.2 An iron-dust-cored inductor and symbol

These plates are called *laminations*, and they make the inductor more efficient — see Section 19.3. The symbol for a laminated iron-cored inductor is also shown in figure 15.3.

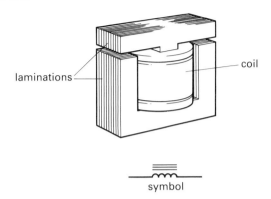

symbol

Figure 15.3 A laminated iton-cored inductor

15.3 *Experiment* B21

Showing the effect of an inductor

Figure 15.4 is a simple circuit for showing the 'choking' effect of an inductor on an alternating current. You will need a laminated iron-core inductor of the type shown in figure 15.3. This inductor should have a value of about 10 henries (10 H) – see Section 17.4.

First you need to close switch SW_1, so that the 6 V d.c. supply lights the lamps.

The variable resistor, *VR*, should be a wire-wound type (see Section 3.5). It is adjusted so that the lamps are about the same brightness. This means that the resistance of the variable resistor is equal to the resistance of the coil of wire in the inductor.

Now open and close the switch. You will see lamp L_2 light up a second or two after lamp L_1. This is because the inductor chokes the build up of current through its coil, and delays the rise of the d.c. current to its steady value.

If you have a 6 V a.c. supply, replace the 6 V d.c. supply by it and you will find that the lamp L_2 never lights. The rapid changes of the a.c. current is so strongly opposed by the inductor that it acts as a high-value resistor in series with lamp L_2. The effect of inductance is explained more fully in Sections 17 and 18.

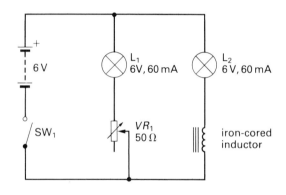

Figure 15.4 Experiment B21: Showing the effect of an inductor

16 The Magnetic Effects of a Current

▽16.1 **Introduction**

It is because an electric current always has a magnetic field associated with it that an inductor resists the flow of a changing current. Many components make use of the magnetic field produced by an electric current, for example the electromagnetic relay described briefly in Book A and later in Chapter 20. Electric motors and generators also make use of the forces which are set up when a coil carrying a current is free to rotate in a magnetic field. Moving-coil meters, such as the multimeter described in Chapter 9, Book A, are also based on the 'motor effect'; that is, a current-carrying coil rotating in a magnetic field.

As you know, a permanent magnet, such as the one used to operate a reed switch (see Chapter 5, Book A) produces a magnetic field, but the following experiments will show you some of the properties of the magnetic field produced by an electric current.

Questions

1 What is meant by the 'field' caused by a magnet?
2 Why is the north pole of a magnet so called?
△ 3 How does a magnetic compass work?

▽16.2 *Experiment* B22

Showing the magnetic effect of a current

A hole should be made in the centre of a piece of stiff card so that a wire, in the simple series circuit shown in figure 16.1(a), can pass through it. At least one small compass should be used to

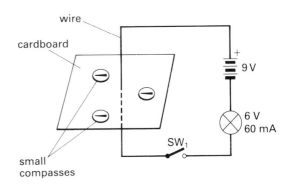

Figure 16.1 (a) Experiment B22: Showing the magnetic effect of a current

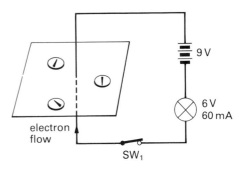

Figure 16.1 (b) The effect on the compasses when current flows through the wire

investigate the shape of the magnetic field around the wire. When no current flows through the wire, as figure 16.1(a) shows, the compass needle will point in the direction of the Earth's magnetic field. In fact, the needle tries to set itself along the field lines due to the Earth's magnetic field. In Britain, these lines actually enter the Earth at an angle of about 60 degrees to the horizontal; however, the compass needle cannot set itself in this direction because it cannot move about a horizontal axis.

Now switch on the current through the wire as shown in figure 16.1(b), Electrons will now flow up through the wire, and you will notice that the needle of the compass will move and set itself as shown.

Questions

1   Move the compass or compasses around the wire. What do you think is the shape of the magnetic field?

2   Reverse the battery connections so that the electrons flow down through the wire. What change does this make to the setting of the compasses?

Remember

A magnetic field is always produced when electrons move. Electrons moving along a straight wire give rise to a magnetic field consisting of concentric circles, as shown in figure 16.2.

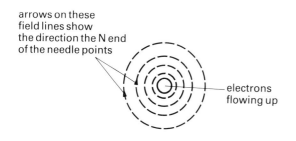

arrows on these field lines show the direction the N end of the needle points

electrons flowing up

Figure 16.2 The shape of the field lines around a current-carrying wire

▽ 16.3 *Experiment* B23

Investigating the magnetic field produced by a solenoid

Wind about 20 turns of insulated copper wire (enamelled, PVC or cotton-covered) onto a cardboard tube as shown in figure 16.3. Use the compass to investigate the magnetic field produced by this coil (which is called a *solenoid*). You will find that quite a strong magnetic field is produced along the axis of the coil and it is strongest in the middle of the coil. In fact, when an easily magnetised material such as 'soft' iron is

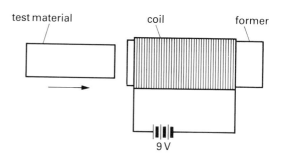

test material   coil   former

9 V

Figure 16.3 Experiment B23: Investigating the magnetic field produced by a solenoid

placed inside this coil, an *electromagnet* is obtained which produces a very much stronger magnetic field, even though the current is unaltered.

Questions

1   Investigate the effect of placing different materials inside the solenoid on the strength of the magnetic field.

2   One end of the solenoid acts as if it were the north pole of a magnet and the other as its south pole. Can you use the compass to find whether this is true?

Figure 16.4 shows the shape of the magnetic field produced by current flowing through the coil.

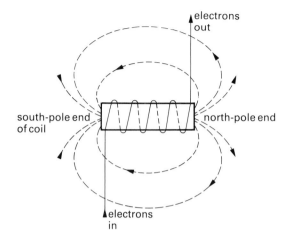

Figure 16.4 The magnetic field produced by a
△ solenoid

16.4 *Experiment* B24

Making a buzzer

A buzzer is a simple little electrical device that makes a noise. The noise is made by the vibrations of a magnetic material as it is drawn towards and then released from one pole of an electromagnet. One way of making a simple buzzer is shown in figure 16.5, overleaf.

You will need an iron or steel bolt to form the magnetic core of the solenoid. Sizes are not critical, but the bolt should be about 30 mm long and between, say 6 mm and 1.5 mm in diameter. You don't need a nut. A piece of scrap wood about 100 mm by 50 mm and 15 mm thick should do as the 'breadboard' for the buzzer. Near one end of the wood, drill a hole part of the way through, which is just smaller than the

diameter of the bolt. Screw the bolt firmly down into the hole.

Cover the threaded portion of the bolt with sellotape or insulating tape. On this 'former', wind about 400 to 500 turns of 0.6 mm diameter (approximately) of enamelled copper wire. Cover the coil of wire with more tape to hold the turns in place. Leave about 100 mm of free wire at the ends of the coil and clean both ends of the wire for about 15 mm.

Two pieces of tinplate need to be cut: one strip should be about 20 mm wide and 100 mm long and the smaller piece about 50 mm long and 10 mm wide. The larger piece of tinplate should be bent Z-shaped as shown in figure 16.5. Place one of the free ends of the wire under the lower part of the 'Z' and screw it down firmly to the wood to hold the wire in place. The top part of the 'Z' should now cover the top of the bolt with a small clearance of a couple of millimetres.

Now you need to make an electrical contact from the second piece of tinplate. Cut a V-shaped point in one end of the smaller piece of tinplate and bend it into a Z-shape like you did for the larger piece of tinplate. Trap a piece of bare wire under the lower part of the 'Z' and screw this end to the wood a small distance away from the larger piece of tinplate. In this position, the pointed end should just be touching the larger piece of tinplate about half-way up.

Now connect the positive terminal of a 4.5 V, 6 V or 9 V battery to the piece of bare wire under the smaller piece of tinplate, and the other terminal to the spare wire from the coil. Bend the smaller piece of tinplate so that the larger piece begins to bounce on and off the contact. You will notice small harmless sparks as the contact makes and breaks. Make adjustments to the position of the contact so that the larger tinplate clatters on the top of the bolt.

Can you see how the buzzer works? Suppose the point of the smaller piece of tinplate is in contact with the larger piece when the battery connection is made. A circuit is made and current freely flows through the coil making it into an electromagnet. The tinplate over the top of the bolt is magnetised and is attracted downwards. This breaks the contact and current stops flowing through the coil. The springiness lifts the tinplate off the top of the bolt, electrical contact is remade and the movement is repeated. You can experiment with this basic design for a buzzer for best effects. The effectiveness of the buzzer is inproved by connecting a 1N4002 diode across the terminals of the coil as shown.

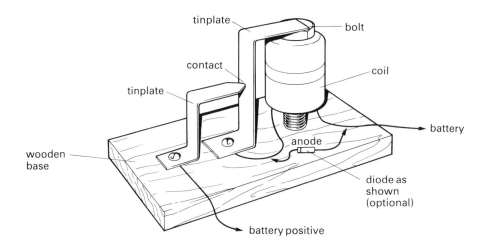

Figure 16.5 Experiment B24: Making a buzzer

16.5   **Loudspeakers**

The magnetic effect of a current is put to good use in the design of loudspeakers. As shown in figure 16.6, when a current, e.g. an audio frequency current from an amplifier, passes through the voice coil, the magnetic field it produces interacts with the loudspeaker's permanent magnetic field and it vibrates in and out between the poles of the magnet. The coil is attached to the core which transmits the movement to a large area of air. The pressure changes produced in the air corresponds to the sound wave which produced the electrical signal in the first place.

Speakers like the one shown in figure 16.7 are usually housed in a speaker cabinet to increase the efficiency with which they move the air in front of them. The design of speaker cabinets is complex. A speaker designed to function at the low end of the audio frequency spectrum, i.e. at around 25 Hz, does not do so well at 16 kHz. Therefore, it is quite usual to find two speakers in one cabinet: one, the *woofer* is designed to operate at low audio frequencies, i.e. below about 3 kHz; frequencies above 3 kHz are handled by a *tweeter*. There may be a mid-range speaker too!

A moving coil loudspeaker is an example of a transducer since it converts energy from one form. i.e. electrical, to another form, i.e. sound. It is quite possible for the moving coil loudspeaker to work the opposite way, i.e. to convert sound energy into electrical energy. That makes it act as a microphone – see Section 17.5.

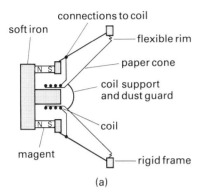

(a)

(b)

Figure 16.6(a) Section through the moving coil loudspeaker

Figure 16.6(b) Symbol for the moving coil loudspeaker

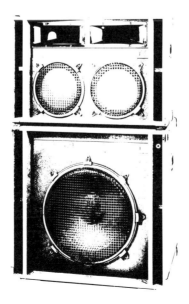

Figure 16.7 Powerful loudspeakers Courtesy: Carlsbro

17 How Magnetism Produces Electricity

17.1 **Introduction**

Experiment B22 in Chapter 16 shows that a magnetic field surrounds a wire through which a current flows. This effect is used to advantage in electric motors, relays, some types of microphones, and other devices.

The magnetic field produced by a current, or by a permanent magnet, can be used to *induce* a current in a nearby conductor. This effect is put to good use in the transformer and electric generator, and it is the reason why an inductor opposes a change of current through it. The two experiments described below explain how these effects come about.

17.2 *Experiment* B25

Producing electricity from magnetism

Wind about 20 turns of insulated copper wire (e.g. 28 or 30 s.w.g. enamelled) on to a cardboard tube as shown in figure 17.1.

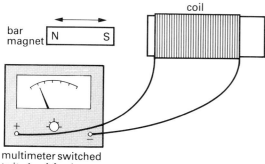

Figure 17.1 Experiment B25: Producing electricity from magnetism

Connect the ends of the wire to a multimeter, switched to its 1 mA f.s.d. range. A lower current range may need to be used.

Bring one pole of a strong bar magnet up to and inside the coil, and note any deflection on the meter.

Questions

1 Bring the other pole of the magnet up to the coil. Which way is the needle deflected?

2 What happens to the deflection when the magnet is drawn out of the coil?

3 Keep the magnet still and move the coil. Do you notice any difference in the deflection?

Carry on experimenting until you are quite sure that electricity flows through the coil only when the magnetic field linked with the coil changes. The changing magnetic field is said to *induce* current in the coil.

17.3 *Experiment* B26

Showing the effect of self-inductance

The circuit shown in figure 17.2 shows that an inductor opposes the growth of current flowing through it. This opposition to a change of current through an inductor is known as *self-inductance*.

The inductor, *L*, should have a value of

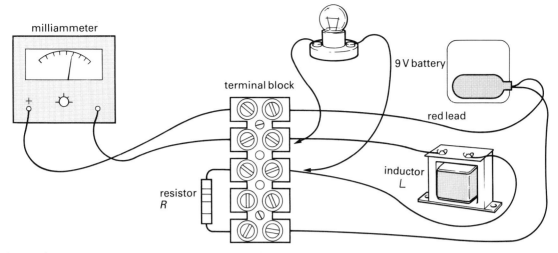

Figure 17.2 Experiment B26: Showing the effect of self-inductance

about one henry. Try the primary or secondary winding of a large mains transformer.

The ammeter should be switched to a range of about 100 mA and the resistor R should have a value of about 100 Ω. The lamp should be left disconnected at first. Press the connector on the battery and current will flow through the series circuit made up of R, L and the ammeter.

Watch the meter carefully. It will take a few seconds to reach a steady value because the self-inductance of the inductor opposes the growing current through it. The reason for the inductor behaving in this way is that a changing current produces a changing magnetic field, and this changing field interacts with the coil of the inductor itself. An induced current flows through the coil which opposes the change of current which produced the field. The voltage which produces the induced current is known as *back e.m.f.* An inductor is said to have a *self-inductance* if it opposes changing current in this way.

Remember

An induced current flows in a direction which opposes the effects producing the current. This statement is often known as *Lenz's law*.

The back e.m.f. generated in a coil will be quite large if the coil has a large inductance, especially if the change of current is rapid. Connect a 6 V, 60 mA lamp across the inductor as indicated in figure 15.3. Allow the current to reach a maximum value and then unclip the battery. You should notice that the lamp flashes briefly. The energy which causes this flash of light comes from the energy of the magnetic field which suddenly collapses and generates a large back e.m.f. This high back e.m.f. is sometimes a problem in circuits — see Section 20.4.

▽ 17.4 **Units for measuring inductance**

The unit of inductance, the *henry* (symbol, H), can be defined in terms of the back e.m.f. One henry is the self-inductance of a coil in which a back e.m.f. of 1 volt is set up by a changing current of 1 ampere per second.

The henry is a rather large unit and, although an inductor of the type shown in figure 15.3 might have a value of a few henrys, most inductors have values of millihenrys and microhenrys. The greater the value of an inductance, the greater its ability to resist the flow of alternating current.

Questions

1   What fraction of a henry is a millihenry?
2   What fraction of a henry is a microhenry?
3   Write down the following inductance values as millihenrys or microhenrys: 0.003 H; 0.000 005 H; 0.02 H.

△

Figure 17.3 Wind energy to electrical energy: an electrical generator at work

18 Time Constant of an Inductive Circuit

▽ 18.1 Reminder

Having studied Chapter 10, you will know that the time constant T of a circuit in which a capacitor charges or discharges through a resistor R is given by

$$T = CR$$

where T is in seconds if R is in ohms and C in farads.

You will remember that T is the time for the charging current to fall by 2/3 of its starting value, or for the voltage across the capacitor to rise to 2/3 of its final value.

Since inductors exert a control on a changing current, a similar time constant △ can be defined for an inductive circuit.

▽ 18.2 Increasing current in an L-R series circuit

The circuit of figure 18.1 (overleaf) shows an inductor L in series with a resistor R. You are to assume that the resistance of L, as measured with an ohm meter, is negligible compared with the value of R.

When the switch SW_1 is moved from position 2 to position 1, a rising current flows through R as shown in figure 18.1(a). The time taken for this current to increase from zero to a value of 2/3 its maximum is the time constant T given by

$$T = L/R$$

where T is in seconds if L is in henrys and R is in ohms.

Questions

1 What is the time constant of an L-R series circuit if $L = 0.5$ H and $R = 1.2$ kΩ?

2 What is the time constant in milliseconds (ms) of an L-R series circuit if $L = 100$ mH and $R = 50$ kΩ?

Figure 18.1(b) shows the growths of voltage across R. This voltage reaches 2/3 of its maximum value (when the maximum △ current flows) in a time of L/R seconds.

▽ 18.3 Decreasing current in an L-R series circuit

If the switch SW_1 is moved to position 2, the current falls to zero. But this fall, or decay, of current is opposed by the inductor. The current falls to 2/3 of its maximum value in a time L/R. Similarly, the back e.m.f. across L falls as shown in figure 18.1(c), but this time the e.m.f. is in the reverse direction. The voltage across R varies as shown in figure 18.1(b), except that once again the voltage is in the reverse △ direction.

▽ 18.4 *Experiment* B27

Showing the effect of inductive reactance

Inductors are used in circuits through which the current changes. In alternating-current circuits, the regular to and fro of the current is opposed by the inductor, which is said to have an *inductive reactance*.

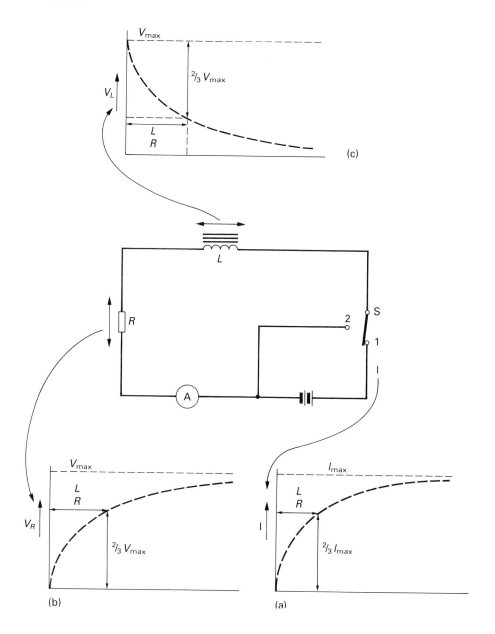

Figure 18.1 Rising current in an *L–R* series circuit

Using the circuit shown in figure 18.2 you can show how this inductive reactance changes with the frequency of an alternating current. The inductor could be the primary or secondary winding of a mains transformer, and the 100 mA meter should be an a.c. type. A waveform generator is used to give a sinusoidal alternating current whose frequency can be varied.

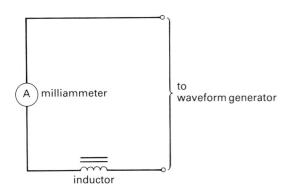

Figure 18.2 Experiment B27: Showing the effect of inductive reactance

Question

1   What happens to the current when the frequency increases from 50 Hz to 500 Hz?

The resistance an inductor presents to an alternating current of frequency f Hz is known as inductive reactance, X_L. It is given by

$$X_L = 2\pi\, fL$$

and is measured in ohms if f is in hertz (Hz) and L is in henrys (H). Your observations with the circuit should confirm this equation: that the inductive reactance increases as frequency increases, shown by a decrease in current.

Questions

2   How does capacitive reactance vary with frequency?

3   Calculate the inductive reactance of a 20 mH inductor at frequencies of 50 and 1000 Hz.

4   What is the inductance of a coil having an inductive reactance of 500 kΩ at 50 Hz?

Note that the inductor always has a resistance — the value given by an ohmmeter. The combination of this 'pure', or d.c. resistance and the inductive reactance gives the inductor's *impedance*. This impedance is not found by adding together the d.c. resistance and the reactance, and an advanced electronics textbook will be needed to explain how it is
△ calculated.

19 Transformers

19.1 **What a transformer does**

The purpose of a transformer is to change an alternating current from one value to another. To increase the voltage an a.c. supply requires the use of a *step-up transformer*; to decrease the a.c. voltage requires a *step-down transformer*.

Transformers are used in low-voltage power supplies which operate from the a.c. mains as explained in Chapter 9, Book C. In this case, a step-down transformer is used.

Transformers are also used to isolate one circuit from another and need not change the voltage. The purpose of an *isolating transformer* is to increase the safety of a.c.

supply by isolating it from the source.

Transformers make use of electromagnetic induction to transfer electrical energy from one coil to another. A changing current through one coil (known as the *primary coil*) induces a current to flow in a nearby coil (the *secondary coil*) as explained below.

19.2 **How a transformer works**

Figure 19.1 shows how a changing magnetic field produced by the changing current in

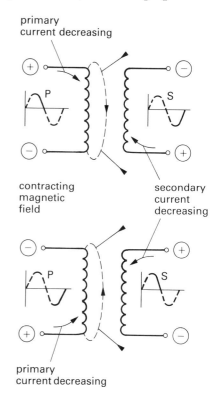

Figure 19.1 How a transformer transfers energy from the primary to the secondary coil

the primary coil induces a changing current in the secondary winding. The graphs **P** and **S** show that the voltage induced in the secondary coil is at any instant of opposite polarity to the primary voltage. This is an important matter in a detailed study of transformer action, and in this course we can say only that there is *mutual inductance* between the two windings of the transformer: current flowing through one winding induces in the other a current whose direction is explained by *Lenz's law of electromagnetic induction* — see Section 17.3.

Note the symbol for a transformer shown in figure 19.2. The meaning of the three lines is explained in Section 19.3.

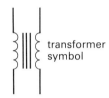

Figure 19.2 Transformer symbol

Questions

1 Why are copper or aluminium not suitable materials for a transformer core?

2 Examine a transformer of the type shown in figure 19.3. How are the primary and secondary windings arranged on the core? Why does it consist of closely-spaced sheets?

▽ 19.3 **Why a transformer core is laminated**

When you examine a low-frequency transformer, such as a mains transformer (50 Hz) or an audio-frequency type (up to 20 Hz), you will notice that iron appears to be the material used for the core and that this core is *laminated*.

Questions

1 How would you show that the core material is probably iron?

2 What does *'laminated'* mean?

The laminations of a transformer are the flat plates of iron (called *stampings*) which are insulated from each other so that there is no electrical connection between the plates.

The laminations are designed to increase the efficiency of a transformer by reducing the heat generated in the core of the transformer. This heat is produced by currents flowing in the core which are induced by the currents flowing in the windings. These currents are called *eddy currents*, since they flow in circles through the iron, as figure 19.4 shows.

In order to reduce the size of these eddy currents, the core is laminated so that the plates lie at right angles to the plane of the coil windings.

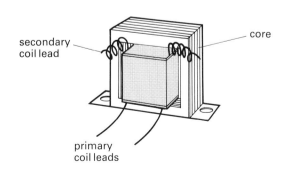

Figure 19.3 The appearance of an audio-frequency transformer

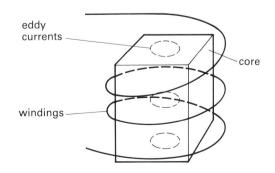

Figure 19.4 The flow of eddy currents in the core

Questions

3   Examine a mains transformer and check that the plates lie along the axes of the windings.

4   Figure 19.5 shows a laminated cube of iron. How would coils be wound on this to make a transformer?

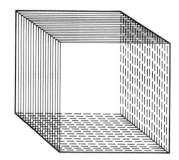

Figure 19.5 A laminated cube of iron

5   Why does heat generated in the core of a transformer represent loss of energy and therefore an inefficient transformer?

6   Why are some large transformers oil-cooled?

▽ 19.4   The transformer equation

Provided a transformer is efficient and transfers electrical energy from one winding to the other with little loss of energy (mostly as heat), a simple equation can be used to calculate the output voltage and the turns ratio of the transformer. In fact, most transformers are almost 100% efficient, and for these the following equation is true:

$$\frac{\text{primary a.c. voltage}}{\text{secondary d.c. voltage}} = \frac{\text{turns on primary}}{\text{turns on secondary}}$$

To see how this equation is used, look at figure 19.6. It shows a step-down transformer and gives 12 V a.c. across the secondary winding with 240 V across the primary coil. These are 1000 turns on the primary coil. How many turns are there on the secondary?

Using the transformer equation,

$$\frac{240}{12} = \frac{1000}{\text{turns on secondary}}$$

To maintain the same ratio 240/12 or 20, the number of secondary turns is 50.

Question

1   Figure 19.7 shows a step-down mains transformer having two secondary windings. Work out the voltages across the secondary windings.

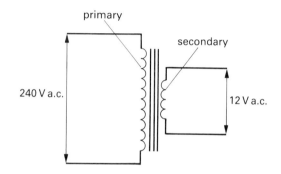

Figure 19.6 Transformer windings for a step-down transformer

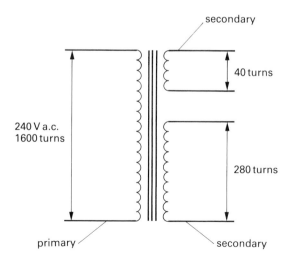

Figure 19.7 A mains transformer with two secondary windings

Note that, in order to work out the secondary voltage, it is not necessary to know the actual number of turns on the primary and secondary windings. It is necessary to know only the *turns ratio:*

$$\text{turns ratio} = \frac{\text{number of turns on primary winding}}{\text{number of turns on secondary winding}}$$

For instance, in the transformer of figure 19.7 you would have obtained the same answers if the primary winding had 800 turns and the secondary windings 20 and 140 turns. This would still make the turns ratios 40/1 (800/20) and 40/7 (800/140).

For a transformer which is 100% efficient, there is no electrical power lost between the primary and the secondary windings. If this is the case, the currents in the windings are in the inverse ratio of the voltages across the windings. To see what this really means, look at figure 19.6. Here the turns ratio is 20/1, the same as the voltage ratio (240/12). Thus the current in the secondary is 20 times the current in the primary, and if 0.5 A flows in the primary, 20 × 0.5 A flows in the secondary.

Questions

2 If an a.c. voltage is stepped up by a transformer, what happens to the current?

3 Assuming no transformer losses and a primary current of 0.2 A, what current flows in each of the secondary windings of figure 19.7?

△

▽ 19.5 **Inductor and transformer core materials**

Air core

Inductors with an air core have a low inductance and are used as high-frequency *chokes* — meaning that they will resist the flow of high-frequency currents. Sometimes they serve this purpose when they consist of just a few turns of wire.

Coils for use at about 1 MHz (a million hertz) use Litz wire, which is made up of a number of fine wires insulated from each other. The purpose of this is to reduce heat dissipation in the wire due to the *skin effect*; this is the tendency for high-frequency currents to flow near the surface of the wire. Thus, the more wires there are, the more surface there is for the current to flow along, and the average current in all the surfaces is lower.

At frequencies higher than 1 MHz, Litz wire is no longer suitable, and thick copper wire, often silver-plated, is used.

Dust and ferrite cores

At frequencies of a few MHz, cores made of ferrite may be used to increase the magnetic field strength and hence the inductance of an inductor.

Dust cores are made of compacted iron dust in wax or some other insulating material. Ferrite is ferric oxide in combination with one or two metal oxides, such as nickel or zinc oxides, and is usually used as the core of the aerial coil in a radio receiver and as the adjustable threaded core in radio-frequency tuned circuits —see figure 15.2. Figure 19.8 (overleaf) shows a selection of ferrite cores.

Questions

1 Examine the inside of a transistor radio and note two ways in which ferrite is used as core material for coils.

2 Use a multimeter to measure the resistance of a rod of ferrite. Compare this with the resistance of a similar rod of iron. Why is there a difference, do you think?

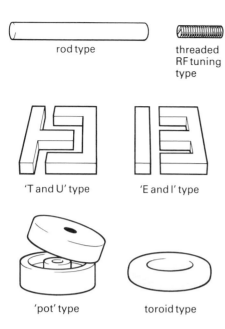

Figure 19.8 A selection of ferrite cores

Ferromagnetic core materials

At frequencies in the audio range (less than 20 kHz), iron is used as the core material for transformers. As explained in Section 19.3 the iron cores of transformers and inductors are laminated so as to improve their efficiency.

Silicon steels and nickel-iron alloys such as *mumetal* and *permalloy* are often used in audio-frequency transformers and inductors since they produce a high magnetic field for a given current in the windings. (They are said to have a high *permeability*.) They also have a high resistivity, hence reducing eddy current losses. Figure 19.9 shows a laminated E-core for use in audio-frequency transformers.

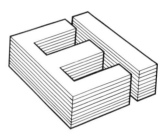

Figure 19.9 A laminated E-core for an audio-frequency transformer

20 Electromagnetic Relays

20.1 What a relay does

As explained in Section 5, Book A, the electromagnetic relay is a type of switch which uses the magnetic effect of a current to open and close contacts. Bear in mind that there are two main purposes of a relay:

(a) To enable a large current to be controlled by a small current. It is therefore a sensitive switch, so that a small current charge can control devices which use heavier currents, such as lamps, motors, solenoids, etc.

(b) To enable the control circuit to be isolated from the controlled circuit.

20.2 The structure of a simple relay

Figure 20.1 shows the essential features of an electromagnetic relay. Note that the controlling current flows through the coil and magnetises the soft iron core of the coil. When the core is magnetised, the soft

iron armature is attracted towards the core when the magnetic pull overcomes the tension in the spring. The movement of the armature *opens*, *closes* or *changes over* the electrical contacts which are in the circuit being controlled.

Notice that there are a large number of turns on the coil of the relay so that a small current flowing through the coil can produce a magnetic field strong enough to attract the armature. Also there is no electrical connection between a circuit connected via the contacts and the current through the coil. It is this complete isolation between the controlled circuit and the control circuit which makes the relay so useful in electronics, especially where low-power equipment such as microcomputers have to operate high-power devices such as lamps and motors.

20.3 The relay symbol and contacts

Figure 20.2 shows the symbol for a relay as it appears on circuit diagrams. The connecting wires to the coil always meet the long sides of the rectangle. The d.c. resistance (for example, 185 Ω) is shown inside the rectangle and by the side of the rectangle is the reference code of the relay. For instance, 'RL$_A$' means first relay, 'RL$_B$' means second relay, etc.

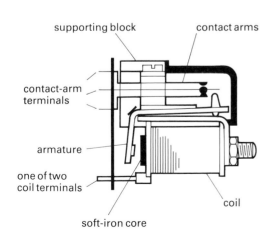

Figure 20.1 The structure of a relay

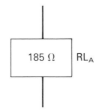

Figure 20.2 Relay symbol

Often this code is underlined to indicate the number of contacts which are used, for example $\underline{RL_C}$ means that the third relay has three contacts in use.
$$3$$

The contacts may be made of silver, tungsten, platinum, or gold. Three kinds of relay contact symbols are shown in figure 20.3. Each symbol indicates the position of the contacts when the relay is not energised, i.e. when no current flows in the coil. Figure 20.3(a) shows a *normally-closed* (NC) contact; figure 20.3(b) a *normally-open* (NO) contact. The code RL_A1 means the first contact of the first relay, RL_B2 the second contact of the second relay, etc. Figure 20.3(c) shows change-over contacts which are common on relays. Note that one pair of contacts breaks before the other makes.

(a)

RL$_A$1

break only (NC)

(b)

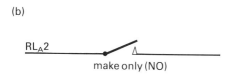

RL$_A$2

make only (NO)

(c)

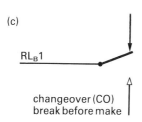

RL$_B$1

changeover (CO)
break before make

Figure 20.3 Types of relay contact

20.4 Using a protective diode in relay circuits

A diode is a device which allows current to flow through it in one direction but not in the opposite direction. Book C explains the properties and uses of the diode in detail, but one important way a diode is used with a relay is described here.

Experiment B27 in Chapter 17 showed that when the energising current through a relay coil falls to zero, a back e.m.f. is generated due to the inductance of the coil. This e.m.f. can be between a few tens of volts and hundreds of volts depending on the type of relay. Such high voltages can damage transistors and integrated circuits which are often used to control the current through the coil of the relay.

As shown in figure 20.4 for a transistor circuit, a diode is connected across the relay coil to protect the transistor. Since the direction of the back e.m.f. is in the opposite direction to the decreasing current which is causing it, the diode has its cathode connected to the positive supply line so that it offers an easy current path for the back e.m.f.

Though you may not be sure of the characteristics of the relay you are using, be sure to use a diode as shown in figure 20.4. The diode can be any silicon type such as one in the series 1N4001 to 1N4006 — see Book C for a discussion of types of diodes and other ways they are used. Book C also describes ways in which a transistor is used to control a relay.

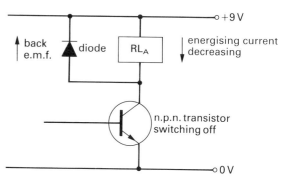

Figure 20.4 The use of a diode to absorb the back e.m.f. when a relay switches off

20.5 *Experiment* B28

Detecting back e.m.f.

The back e.m.f. generated by the coil of a relay can be shown very effectively using the circuit shown in figure 20.5.

You need a single-pole, double-throw relay, i.e. a single-pole changeover type which can be energised by a 9 V battery. The contacts you must identify are the relay's normally-closed contacts. These are connected in series with the 9 V battery, B_1, and the coil, L_1, of the relay. Immediately the circuit is complete, the relay will 'buzz' as the armature rapidly vibrates. The relay first energises then de-energises in rapid succession. You have made a buzzer described in Experiment B24. It works like this:

As the relay contacts are normally closed when the circuit is completed, the relay armature is attracted as current flows through the coil. The normally-closed contracts open, current ceases and the armature springs back. This closes the contacts and the cycle of rapid movement of the armature is repeated.

Now while the relay is buzzing, moisten your finger tips and bridge the coil terminals with them. You will feel the tingle of an electric shock, proof that the relay is generating a voltage way above 9 V caused by the self-inductance of the coil. This back e.m.f. is high enough to light a neon lamp placed across the coil terminals. Look carefully at the two electrodes in the neon lamp; there will be a glow around one of them which shows that a d.c., not an a.c. voltage is being generated. This back e.m.f. is caused by the magnetic field produced by the relay coil rapidly 'collapsing' as the normally-closed contacts open.

Now connect a diode, D_1, e.g. a 1N400I, across the coil as shown. The neon lamp ceases to glow but the buzzer continues to buzz. What has happened to the back e.m.f.? The diode acts as a short circuit to the currents produced by the back e.m.f. which prevents the back e.m.f. from rising more than a few volts.

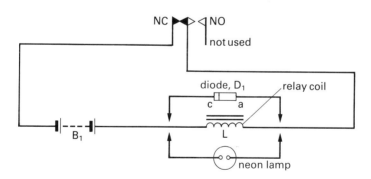

Figure 20.5 Detecting back e.m.f.

21 The Tuned Circuit

▽ ## 21.1 **Resonance**

In a radio circuit (e.g. Project Module B5) a capacitor is connected in parallel with an inductor as shown in figure 21.1(a). This arrangement is known as a *tuned circuit* because it enables different radio stations to be tuned in, usually by adjusting a variable capacitor (*VC*) as shown.

If this circuit is fed with a variable frequency, it will *resonate* at one particular frequency called its *resonant frequency*. When this happens, the voltage generated across the inductor (or capacitor since they are in parallel) is a maximum. Figure 21.1(b) is a graph showing that at the resonant frequency, f_0, of the tuned circuit, the voltage across the circuit rises to a maximum value. The resonant frequency is given by the equation

$$f_0 = 1/(2\pi\sqrt{LC})$$

where L is the value of the inductance (in henries) of the aerial coil and C is the value of the capacitance (in farads) of the tuning capacitor.

Now when this tuned circuit is used in an amplitude modulated (AM) radio it is fed with a wide range of radio waves, called *carrier waves*, from the aerial or aerial coil. For example, on the medium waveband a station which has a wavelength of 300 metres transmits a carrier wave which has a frequency of 1 MHz (one million hertz). If a station happens to have a frequency equal to the resonant frequency of the circuit, its transmissions will cause a high voltage to build up across the tuned circuit. Stations having lower or higher frequencies will not produce such a large voltage. So only the radio transmission which causes the tuned circuit to resonate is detected, for it is the one which needs least amplification to reach the earpiece or loudspeaker.

The tuned circuit depends on the different properties of a capacitor and an inductor. Remember that the reactance of a capacitor decreases with increasing frequency while the reactance of an inductor increases with increasing frequency. When these two properties

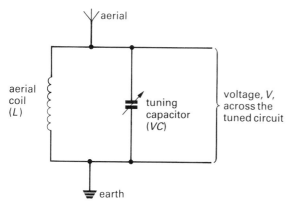

Figure 21.1 (a) A tuned circuit in a radio receiver

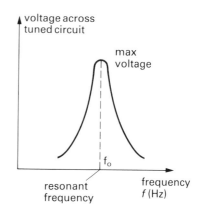

Figure 21.1 (b) The way the voltage across a tuned circuit varies with the frequency

combine in the tuned circuit, a resonant circuit is produced. A complete understanding of the way the parallel tuned circuit works will be found in books more advanced than *Basic Electronics*.

Note that the equation above for the resonant frequency of a tuned circuit indicates some interesting facts about the values of the inductor and capacitor in a radio receiver. For instance if the value of either C or L (or both) is increased, the frequency which is tuned in is increased, i.e. radio transmissions of shorter wavelengths are tuned in. Thus the tuning capacitor has to have its value reduced to tune from the long to the short wave end of the tuning scale on a radio. So shortwave receivers have tuning coils of a few turns; longwave receivers have tuning coils of many turns. Project Module B5 describes a radio receiver that uses a tuned circuit to select radio stations.

21.2 **Selectivity**

At the resonant frequency, f_0, the voltage developed across the terminals, T, of the tuned circuit is at a maximum value. It is this voltage which the radio produces strongly and is heard as the signal from the station transmitting on a frequency f_0.

However, the tuned circuit does not select just one frequency but a band of frequencies as shown in figure 22.1. The *bandwidth* is defined as the difference between f_h and f_l, the difference between the high and lower frequencies on either side of the resonant frequency. These two frequencies are defined as the frequencies where the voltage has fallen to 0.7 of the maximum voltage.

$$\text{bandwidth} = f_h - f_l$$

Obviously, if the tuned circuit is able to separate two stations which are transmitting at frequencies which are close together, the bandwidth must be small — but not too small, since the carrier wave from the station requires a certain minimum bandwidth in order to carry a full range of audio frequencies.

The quantity Q defined by the equation:

$$Q = f_0/(f_h - f_l)$$

is a measure of the sharpness of resonance of the *L-C* parallel resonant circuit. It is known as the quality factor of the circuit, or just simply the 'Q' of the circuit. Large values of Q give rapid rates of fall of voltage on either side of the maximum voltage, and hence correspond to sharp tuning of the circuit.

Figure 21.2 Radio receiver design has improved dramatically during the 1900s. This model is for shortwave use Courtesy: Sony (UK) Ltd.

22 Project Modules

22.1 What they are

At the end of each book of *Basic Electronics* there are a number of practical projects for you to build. These projects are called Project Modules and there are thirty five of them in all. This chapter describes how to build and use the seven Project Modules shown in figure 22.1. They are:

B1: Schmitt Trigger
B2: Relay Driver
B3: Bistable
B4: Monostable
B5: Radio Receiver
B6: Infrared Source and Sensor
B7: Metal Detector

The Project Modules enable you to build up a set of electronic building blocks which can be connected together in various ways to design useful and interesting electronic systems. Details are provided for assembling each Project Module on a printed circuit board (PCB) and for interconnecting it with other Project Modules using flying leads.

Before assembling the circuits, you should read Book A, Section 6.3, which gives guidance on the preparation of the PCBs. You should also read Book A, Section 12.2, which gives hints on handling the CMOS devices used in three of these projects. Examples are given of useful electronic systems which can be designed using the Project Modules.

Figure 22.1 The first three Project Modules described in this chapter

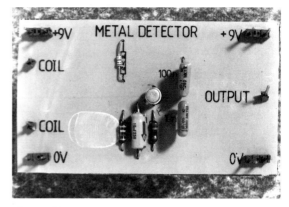

Figure 22.1 The last four Project Modules described in this chapter (IR Source and IR Sensor make one Project Module together)

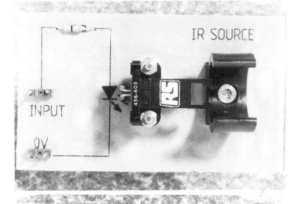

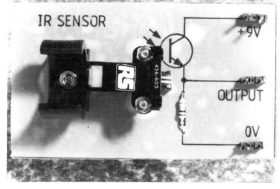

22.2 *Project Module* B1

Schmitt Trigger

What it does

The Schmitt Trigger converts a slowly changing voltage to a sharply changing voltage. It is put to good use in circuits where changes of light intensity, temperature, or pressure need to activate or 'trigger' electronic switching, counting and control circuits. Book D, Chapter 4 describes how the Schmitt Trigger works, and Book E, Chapter 9 gives further details of the integrated circuit used in this project.

Circuit

Only one of the four Schmitt triggers in the integrated circuit package, IC_1, is used in the circuit shown in figure 22.2 (overleaf). This CMOS digital IC is also used as the basis of the Pulser (Project Module A2). Note that a voltage divider, made up of VR_1 and LDR_1, is connected to the input of IC_1. As the illumination of LDR_1 varies, the voltage at point X rises and falls smoothly. The Schmitt Trigger responds to two values of this voltage and, at each value, produces a sharp change in the output voltage. Thus a falling voltage on the input causes the output to switch to a high state; a rising voltage causes it to switch to a low state. The high and low states of the output are registered by LED_1, i.e. if LED_1 is off, the output is low, and vice versa.

Components and materials

IC_1: quad 2–input NAND Schmitt Trigger CMOS type 4093
IC holder: 14–way
VR_1: linear or log variable resistor, value 100 kΩ

LDR_1: light dependent resistor, type ORP12 or similar
R_1: fixed-value resistor, value 1 kΩ
LED_1: light emitting diode
terminal block: 5-way length
battery: PP9
battery clip: PP9 type
wire: multistrand, e.g. 7 × 0.2 mm; single strand: e.g. 1 × 0.6 mm
nylon screws: 2 × 6BA
connectors: PCB header and PCB socket housing; crimp terminals

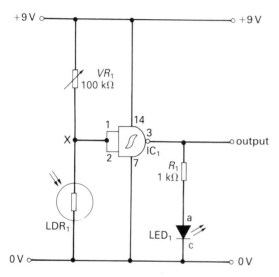

Figure 22.2 Circuit diagram of the Schmitt Trigger

Printed circuit board (PCB) assembly

Figure 22.3 shows the layout of the components on the PCB, and figure 22.4 the copper track pattern on the other side of the PCB. See Book A, Section 6.3, for guidance on the preparation of the PCB.

Make sure that LED_1 is connected the right way round with its cathode terminal nearest the edge of the board. The 6BA nylon screws are used to fix the terminal block to the PCB. Note that four link wires are needed from the terminal block to the PCB. Use a small screwdriver to fix LDR_1 and VR_1 to the terminal block, and check all soldered joints carefully before connecting the Schmitt Trigger to the battery.

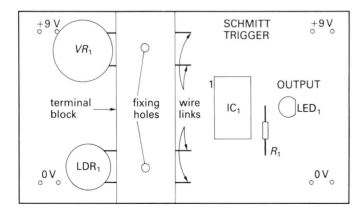

Figure 22.3 Component layout of the PCB (actual size)

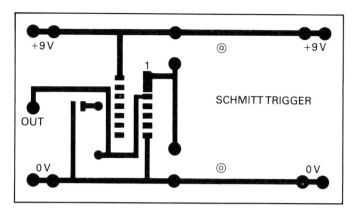

+9 V

+9 V

SCHMITT TRIGGER

OUT

0 V

0 V

Figure 22.4 Track pattern for the PCB (actual size)

The terminal pins for the connections on the PCB are made by cutting single and double pins from the PCB header. Single sections of the PCB socket housing are used for terminating wire ends for connecting the Schmitt Trigger to the battery and to other Project Modules. Strip 5 mm of insulation from the ends of the wires, and use a crimping tool to squeeze a crimp connector on the bare ends. Push the crimp connector into the PCB socket until it clicks into place.

Testing and use

Adjust VR_1 so that LED_1 just comes on. Cover LDR_1 and LED_1 will go off. Uncover it and it will come on again. Thus the Schmitt Trigger acts as an electronic switch. Changes in light intensity detected by LDR_1 produce changes in the input voltage at the input of IC_1 which makes the output voltage switch between the two states high (LED_1 on), and low (LED_2 off).

Note that LDR_1 can be replaced by a thermistor or reed switch if the switching is to be activated by heat or a magnet, respectively. The value of VR_1 might need to be changed if a thermistor is used.

Connect the output of the Schmitt Trigger to the input of the Triple Lamp (Project Module A7) so that a brighter 'high' output is obtained when the Schmitt Trigger operates. Figure 22.5 (overleaf) shows some options for using the Schmitt Trigger with other Project Modules.

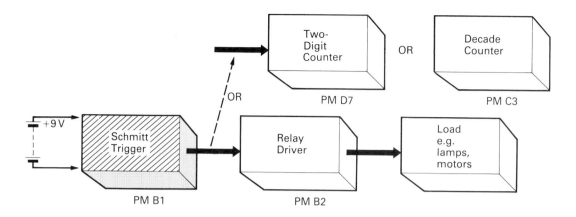

Figure 22.5 Some options for using the Schmitt Trigger with other Project Modules

22.3 *Project Module* B2

Relay Driver

What it does

This module enables low power signals, e.g. those from the Schmitt Trigger (Project Module B1), to switch on and off higher power devices such as lamps and motors. The two single-pole double-throw relays on this Project Module provide independent on-off control of two motors.

Circuit

Figure 22.6 shows how two field-effect transistors (FETs), Tr_1 and Tr_2, are used to energise the relays. The particular advantage of using a VMOS — **v**ertical **m**etal-**o**xide **s**emiconductor — transistor in the Relay Driver is that its gate terminal, g, draws negligible current from the signal source for it to control a large current flowing between its drain, d, and source, s, terminals. Thus the FET is voltage-operated, not current-operated as with bipolar transistors. It is an ideal device for amplifying the low-power signals from CMOS devices, e.g. the Schmitt Trigger and Pulser. The properties of the FET are

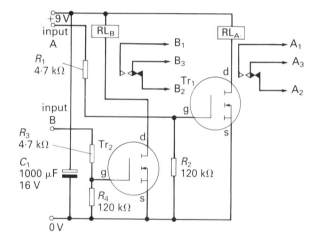

Figure 22.6 Circuit diagram of the Relay Driver

described in Book C, Chapter 27, and the relay in Chapter 20 of this book.

The changeover contacts on relays RL_A and RL_B are able to switch a current of 1 A which is enough to control small d.c. motors. When a relay energises, two normally-open (NO) contacts, e.g. A_1 and A_3, close, and two normally-closed (NC) contacts, e.g. B_3 and B_2, open. Contacts A_3 and B_3 are common on each relay. The large-value electrolytic capacitor, C_1, stabilises the supply voltage to the circuit.

Components and materials

Tr_1, Tr_2: n-channel field-effect transistor, type VN46AF

R_1 to R_4: fixed-value resistors, 0.25 W, ±5%

RL_A, RL_B: subminiature single-pole changeover relays; NB check that the pin layout on the relays you use is compatible with the PCB pattern of figure 22.3.

C_1: electrolytic capacitor, value 1000 μF 16 V

battery: 9 V PP9

battery clip: PP9 type

wire: multistrand, e.g. 7 × 0.2 mm

PCB: 90 mm × 50 mm

connectors: PCB header and PCB socket housing; crimp terminals

PCB assembly

Figure 22.7 shows the layout of the components on the PCB, and figure 22.8 the copper track pattern on the other side to the PCB. See Book A, Section 6.3, for guidance on the preparation of the PCB.

The terminal pins for the connections on the PCB are made by cutting single and double pins from the PCB header. Single sections of the PCB socket housing are used for terminating wire ends for connecting the Relay Driver to the battery and to other Project Modules. Strip 5 mm of insulation from the ends of the wires, and use a crimping tool to squeeze a crimp connector on the bare ends. Push the crimp connector into the PCB socket until it clicks into place.

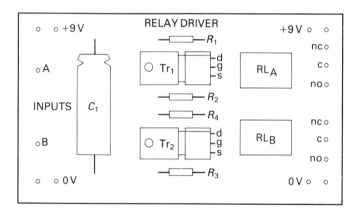

Figure 22.7 Component layout on the PCB (actual size)

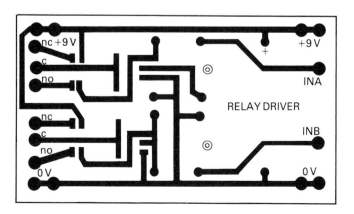

Figure 22.8 Track pattern for the PCB (actual size)

Testing and use

Connect the power supply pins of the Relay Driver to those on the Schmitt Trigger — Project Module B1). Connect either input A or B on the Relay Driver to the output of the Schmitt Trigger. Adjust VR_1 on the latter so that LED_1 just goes off when LDR_1 is covered. One of the relays will 'click' as it energises and de-energises when LDR_1 is covered and uncovered. Check that relay RL_B also works by connecting input B to the Schmitt Trigger.

Figure 22.9 is a block diagram of a light-operated control system. It switches the motor on and off when LDR_1 is covered and uncovered. Note that the motor must be operated by a separate power supply, not the one used for activating the circuits.

Modify this simple on/off control system to produce the following results:

(a)   Wire up the d.c. motor to the Relay Driver so that on covering LDR_1 the motor starts, and stops when it is uncovered.

(b)   Wire up the d.c. motor to the Relay Driver so that on uncovering LDR_1 the motor starts, and stops when it is covered.

(c)   Connect the two Project Modules to a small d.c. motor-driven vehicle so that the vehicle can be controlled remotely by a torch.

(d)   Switch on a low-voltage lamp with the system *but do not connect mains-operated equipment to the system.*

(e)   Use the Pulser (Project Module A2) to switch a d.c. motor on and off at regular intervals.

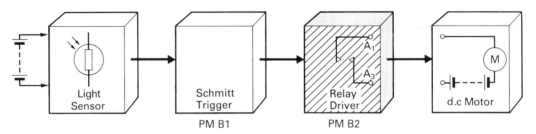

Figure 22.9 Light-operated control system

22.4 *Project Module* B3

Bistable

What it does

This Project Module is used as an electronic switch which 'remembers' whether it is on or off until 'told' to change its state. The bistable (also called a flip-flop) is a digital circuit which stores a single binary digit, and it is the basis of counting and memory circuits described in Book E. The Bistable has two outputs: when one output is high the other is low, and the outputs are switched over by a high-to-low change at the input. The Bistable is used in

the series of Project Modules as a latching on/off switch. For example, a high-to-low signal from the Schmitt Trigger (Project Module B1) energizes the Relay Driver (Project Module B2) until it is switched off by a second high-to-low signal from the Schmitt Trigger.

Circuits

The circuit shown in figure 22.10 is based on an integrated circuit, IC_1. In fact, IC_1 contains two identical bistable circuits, only one of which is used. The bistable used has one input, pin 13, and two outputs,

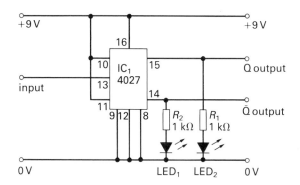

Figure 22.10 Circuit diagram of the Bistable

pins 14 and 15. These two outputs are marked Q and $\bar{Q}$ to indicate that when one output is, say, high, the other is low. The high and low states of the outputs are indicated by the two light emitting diodes, LED_1 and LED_2.

Components and materials

IC_1: dual JK bistable (or flip-flop), CMOS type 4027
IC holder: 16–way
R_1, R_2: fixed-value resistors, 0.25 W, ±5%
LED_1, LED_2: light emitting diodes
battery: 9 V PP9
battery clip: PP9 type
wire: multistrand, e.g. 7 × 0.2 mm
PCB: 90 mm × 50 mm
connectors: PCB header and PCB socket housing; crimp terminals

PCB assembly

Figure 22.11 shows the layout of the components on the PCB, and figure 22.12 the copper track pattern on the other side

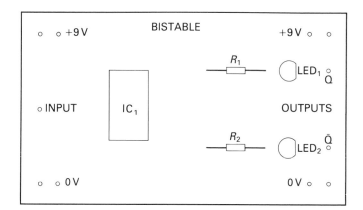

Figure 22.11 Component layout of the PCB (actual size)

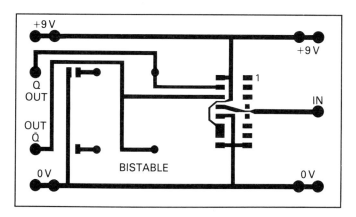

Figure 22.12 Track pattern for the PCB (actual size)

of the PCB. See Book A, Section 6.3, for guidance on the preparation of the PCB.

Make sure that LED_1 and LED_2 are connected the right way round; the cathode terminals of each LED should be nearest the edge of the PCB. The terminal pins for the connections on the PCB are made by cutting single and double pins from the PCB header. Single sections of the PCB socket housing are used for terminating wire ends for connecting the Bistable to the battery and to the other Project Modules. Strip 5 mm of insulation from the ends of the wires, and use a crimping tool to squeeze a crimp connector on the bare ends. Push the crimp connector into the PCB socket until it clicks into place.

Testing and use

Make sure that the Schmitt Trigger (Project Module B1) is working. Connect the power supply connections of the Schmitt Trigger to those on the Bistable. Connect the output pin of the Schmitt Trigger to the input pin of the Bistable.

Adjust VR_1 on the Schmitt Trigger so that the light emitting diode (LED_1) switches on and off when LDR_1 is covered and uncovered. Note how LED_1 and LED_2 on the Bistable behave: if LED_1 is on, LED_2 is off, and they switch on and off alternately as the Schmitt Trigger sends successive signals to the Bistable. The outputs stay in a particular state until the next signal is received from the Schmitt Trigger.

This module can be used in the light-operated control system shown in figure 22.9 to make the relay 'latch on' when LDR_1 senses a change of light intensity. The relay will then switch off when the next signal is received. The Bistable needs to be connected between the Schmitt Trigger and the Relay Driver. Modify the light-operated control system shown in figure 22.9 to produce the following results:

(a) Wire up the d.c. motor to the Relay Driver so that on uncovering LDR_1 on the Schmitt Trigger, the motor starts and continues to run until LDR_1 is covered and uncovered again.

(b) Connect this control system to a small d.c. motor-driven vehicle so that it can be controlled by light from a torch. One flash of light on LDR_1 starts the vehicle, and the next flash stops it.

(c) Remotely switch on a low voltage filament lamp with the Relay Driver *but do not connect mains-operated equipment to the system.*

(d) In what useful ways can you use the Pulser (Project Module A2) with the Bistable?

22.5 *Project Module* B4

Monostable

What it does

This Project Module gives a time delay. This delay is the length of time that the output voltage of the Monostable remains high. The Monostable can be used to provide an alarm, e.g. a buzzer or a flashing light, at the end of a time delay. Or it can be used to control motors for predetermined periods of time. In some circuits, the Monostable is used as a 'pulse stretcher' to lengthen a short pulse so that another circuit can be operated more reliably.

The time delay produced by the Monostable can easily be changed by changing the values of a resistor and/or a capacitor. You will be able to use the Monostable with other Project Modules, e.g. with the Relay Driver (Project Module B2), and to control lamps, motors, etc.

Circuit

The circuit shown in figure 22.13 is based on an integrated circuit, IC_1, a 555 timer. This device is described in Chapter 14. The time, T, that the output pulse from pin 3 remains high after the circuit is 'triggered' is determined by just two components, R_1 and C_1. In this circuit, the output pulses switch on and off, two light emitting diodes, LED_1 and LED_2, which are helpful in seeing whether the circuit is working.

When the power supply is connected to the circuit, the Monostable is in its normal state with the output voltage at pin 3 low, i.e. near 0 V. On pressing switch, SW_1, and releasing it, the Monostable is said to be 'triggered' and the voltage at pin 3 rises to high, i.e. near +9 V. The time for which the output pulse remains high is given by the product of the values of R_1 and C_1, i.e.

$$T = 1.1 R_1 \times C_1$$

For the values of R_1 and C_1 shown in figure 22.14, the time delay is given by

$$T = 100 \times 10^3 \times 100 \times 10^{-6} = 10 \text{ s}$$

LED_1 is on and LED_2 is off before the switch, SW_1, is pressed. LED_2 is on and LED_1 is off during the time delay. At the

end of the time delay, the voltage at pin 3 falls to near zero again and LED_2 goes off and LED_1 comes on again. The Monostable remains in this normally-off state until SW_1 is pressed again.

For some applications it is useful to be able to trigger the Monostable by an external trigger pulse, e.g. from the Schmitt Trigger circuit, rather than by pressing SW_1. The coupling capacitor, C_3, allows the external pulse to be applied to pin 2 without there being any direct current path between the two circuits. The use of a capacitor to couple a varying signal between two different circuits is explained in Chapter 13.

As explained in Chapter 14, pin 4 on the 555 is used to reset the timer to its normally-off state at any time during the time delay. In figure 22.14, capacitor C_2 and resistor R_3 ensure that the 555 automatically sets to its normally-off state when the power supply is switched on.

Components and materials

IC_1: integrated circuit timer, type 555
IC holder: 8–way
R_1 to R_5: fixed-value resistor, 0.25 W, ±5%
LED_1, LED_2: light emitting diodes green and red, respectively

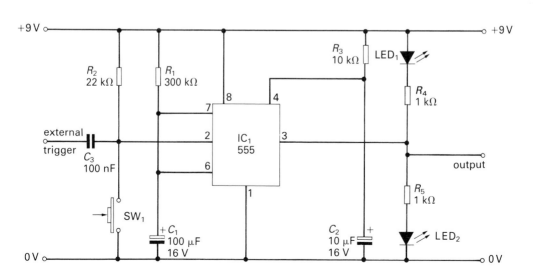

Figure 22.13 Circuit diagram of the Monostable

C_1, C_2: electrolytic capacitors, 16 V
working
C_3: polyester capacitor
terminal block: 5-way length
battery: 9 V PP9
battery clip: PP9 type
wire: multistrand, e.g. 7 × 0.2 mm; single
strand: e.g. 1 × 0.6 mm
SW_1: push-to-make, release-to-break
keyboard switch
PCB: 90 mm × 50 mm
connectors: PCB header and PCB socket
housing; crimp terminals

PCB assembly

Figure 22.14 shows the layout of the
components on the PCB, and figure 22.15
the copper track pattern on the other side
of the PCB. See Book A, Section 6.3, for
guidance on the preparation of the PCB.

Make sure that LED_1 and LED_2 are
connected the right way round. The
cathode of LED_1 faces away from the edge
of the board, the cathode of LED_2 towards
the edge of the board. Use nylon screws to
secure the terminal block to the PCB. Note
that three wire links are needed between
the terminal block and the PCB.

The terminal pins for the connections on
the PCB are made by cutting single and
double pins from the PCB header. Single
sections of the PCB socket housing are
used for terminating wire ends for
connecting the Monostable to the battery
and to other Project Modules. Strip 5 mm
of insulation from the ends of the wires,

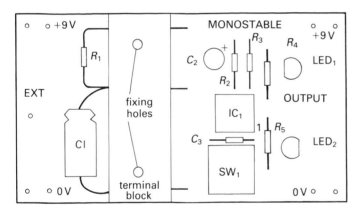

Figure 22.14 Component layout on the PCB (actual size)

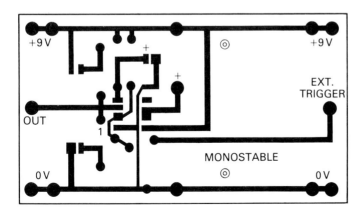

Figure 22.15 Track pattern for the PCB (actual size)

and use a crimping tool to squeeze a crimp connector on the bare ends. Push the crimp connector into the PCB socket until it clicks into place.

Testing and use

Connect the battery and LED_1 (the green LED) will be on and LED_2 (the red LED) off. Press SW_1 and release it, and LED_1 will go off and LED_2 will come on for a time delay of about 10s. Experiment with different values of R_1 and C_1. By replacing R_1 by a variable resistor, a precise time delay can be obtained. If you want to reset the Monostable after a time delay has been started, simply disconnect the power supply and reconnect it.

1 Timer-alarm

Make the Pulser (Project Module A2) produce a high-pitched audio alarm in the Piezo-sounder (Project Module A3). Connect these two Modules to the Monostable as shown in figure 22.17. In this system, the Monostable is used to switch on the audio alarm after a time delay. Before SW_1 is pressed the alarm sounds and it stops sounding as soon as SW_1 is pressed. It sounds again after the time delay $T = R_1 \times C_1$.

Questions

1 How would you connect these three modules together so that the alarm sounds during the time delay?

2 How would you use the Audio Amplifier (Project Module A5) and the Loudspeaker (Project Module A3) so that a louder alarm tone is produced?

2 Motor Controller

How would you use the Monostable with the Relay Driver (Project Module B2) to drive a lamp or a d.c. motor for a preset period of time, or to switch it on after a preset period of time?

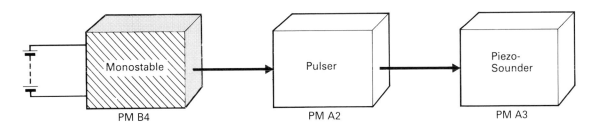

Figure 22.16 Timer alarm system

22.6 Project Module B5

Radio Receiver

What it does

This Project Module receives AM (amplitude modulated) radio broadcasts so that they can be heard in a crystal earpiece. If loudspeaker reception is required, the Audio Amplifier (Project Module A5) and the Loudspeaker (Project Module A3) may be added.

Circuit

The main component in figure 22.17 is the special transistor-like integrated circuit, IC_1, which is the ZN414 'radio on a chip'. IC_1 does not contain the tuned circuit

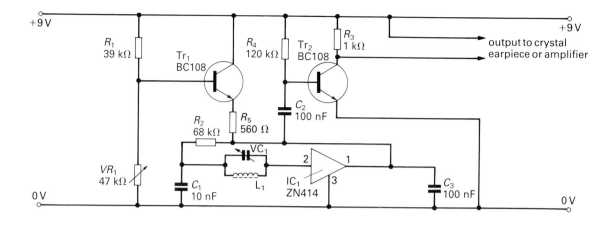

Figure 22.17 Circuit diagram of the Radio Receiver

(variable capacitor, VC_1, and aerial coil, L_1), that all radio receivers need. But it does contain all the circuitry needed to amplify the weak radio frequency signals selected by the tuned circuit. And it produces enough audio power to operate a crystal earpiece for personal listening. Transistor Tr_2 provides an extra stage of amplification to help boost the power for an earpiece. Transistor Tr_1, resistor R_1, and variable resistor VR_1 are used to set the power supply voltage at pin 3 of IC_1 to about 1.3 V. Adjustment of this power supply using VR_1 alters the sensitivity of the circuit. The way a tuned circuit selects a particular radio station is described in Chapter 21.

Components and materials

IC_1: tuned radio frequency amplifier, type ZN414

VR_1: miniature preset variable resistor, value 47 kΩ

R_1 to R_5: fixed-value resistors, values 39 kΩ, 68 kΩ, 120 kΩ, 1 kΩ and 560 Ω, respectively, all 0.25 W, ±5%

VC_1: tuning capacitor, 100 pF maximum value suitable, but 500 pF will do

Tr_1, Tr_2: npn transistors, type BC108 or similar

L_1: medium wave coil, preferably bought but can be made using 0.6 mm enamelled copper wire — see the following text

ferrite rod: length about 50 mm, diameter about 10 mm to fit inside L_1

knob: to fit VC_1

battery: 9 V PP9

battery clip: PP9 type

C_1 to C_3: polyester capacitors, values 10 nF, 100 nF, 100 nF, respectively

wire: multistrand, e.g. 7 × 0.2 mm; single strand, e.g. 1 × 0.6 mm

PCB: 90 mm × 50 mm

connectors: PCB header and PCB socket housing; crimp terminals

PCB assembly

Figure 22.19 (overleaf) shows the layout of the components on the PCB, and figure 22.20(overleaf) the copper track pattern on the other side of the PCB. See Book A, Section 6.3, for guidance on the preparation of the PCB.

The terminal pins for the connections on the PCB are made by cutting single and double pins from the PCB header.

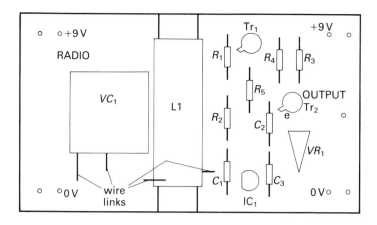

Figure 22.18 Component layout on the PCB (actual size)

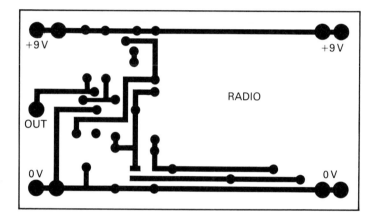

Figure 22.19 Track pattern for the PCB (actual size)

Single sections of the PCB socket housing are used for terminating wire ends for connecting the radio receiver to the battery and to other Project Modules. Strip 5 mm of insulation from the ends of the wires, and use a crimping tool to squeeze a crimp connector on the bare ends. Push the crimp connector into the PCB socket until it clicks into place.

Solder short lengths of wire from VC_1 to the PCB. Special multistranded 'Litz' wire is used on the aerial coils you can buy, which gives better selectivity than a coil wound from single strand enamelled copper wire. If you make the coil yourself, close-wind about 30 turns of 0.315 mm

enamelled copper wire on a cardboard former into which the ferrite rod fits loosely. Make sure that the enamel is scraped off the ends of the wire before soldering the ends into the PCB. It is worth experimenting with the number of turns of wire. Moving the ferrite rod in and out of L_1 makes tuning easier by bringing in different wavebands. Without the ferrite rod in the coil, it is possible to receive shortwave transmissions at night.

Testing and use

Though the Radio Receiver will operate a crystal earpiece, it is best tested by connecting it to the Audio Amplifier (Project Module A5) and Loudspeaker (Project Module A3) as shown in figure 22.20. Adjust the variable resistor on the Audio Amplifier for maximum gain. Connect the 9 V battery to the circuit and adjust the sensitivity control, VR_1, on the Radio Receiver so that the speaker is just on the point of 'whistling' — the maximum sensitivity. Adjust the tuning capacitor to tune in a station. You may need to adjust the position of the ferrite rod in the coil to receive the station you want. And, since the aerial coil has directional sensitivity — it picks up stations best which are in line with the axis of the coil — you may need to turn the Radio Receiver for best effect.

Obviously, the above system gives you a Radio Receiver with loudspeaker reception which is perhaps all you want to do with it.

However you may want to use it as the main 'information provider' in the systems shown in figures 22.22 and 22.23.

1   Remote control of the Radio Receiver
Assemble the system shown in figure 22.22. It enables the Radio Receiver to be switched on and off by a beam of light. One flash of light switches it on, and the next switches it off. Explain the functions of the Modules in this system.

2   Radio timer
Assemble the system shown in figure 22.23. It enables the Radio Receiver to be switched off after a time delay set by the values of capacitor, C_1, and resistor, R_1, in the Monostable. How would you use the Relay Driver to switch on the Radio Receiver after a time delay?

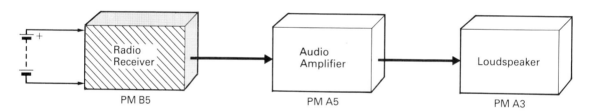

Figure 22.20 Radio receiver system

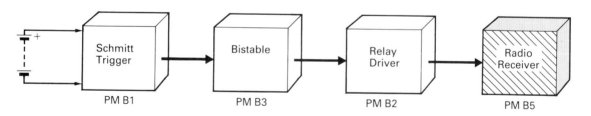

Figure 22.21 System for the remote control of the radio receiver

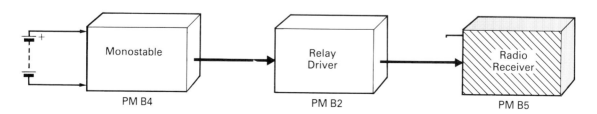

Figure 22.22 System for the radio timer

22.7 *Project Module* B6

Infrared Source and Sensor

What they do

These two Project Modules send and receive an infrared beam which can carry speech and music through a fibre optics cable or through the air using lenses. They enable you to build a communications system for transmitting speech by using two Audio Amplifiers (Project Module A5) and two Loudspeakers (Project Module A3).

Circuits

The two simple circuits required are shown in figure 22.23. The infrared beam is generated by an infrared emitting diode, LED_1, which has a resistor, R_1, connected in series with it to allow it operate safely at its correct operating voltage when connected to the Audio Amplifier. Changes in the signal strength from the Audio Amplifier produce corresponding changes in the strength of the infrared beam generated by LED_1.

This amplitude modulated (AM) infrared beam is detected by a phototransistor, Tr_1, in the receiving circuit. In combination with the resistor, R_1, the small current changes induced in the transistor are converted into small voltage changes at the emitter of Tr_1. These small amplitude signals are passed to the Audio Amplifier which provides an output to the Loudspeaker. The phototransistor is described in Book C, Chapter 11.

PCB assembly

Figure 22.24 (overleaf) shows the PCB track pattern and figure 22.25 (page 107) the component overlay. To make sure LED_1 is about 15 mm above the PCB, it is soldered to the pins of a two-way PCB connector and its leads bent through 90°. LED_1 is pushed into a TO18 'device housing' which is a special type of connector for fibre optics cable. This connector is mounted on stand-off screws that enable LED_1 to slip into the hole provided.

The connector enables a fibre obtics cable to be plugged into place, or the beam of infrared generated by LED_1 can be focused by a lens for free-air transmission. At a height of 15 mm, LED_1 is facing along the axis of a convex lens which is mounted in a tube clipped into a plastic holder that allows the tube to move horizontally and focus the beam. An identical arrangement is used to hold the sensor, Tr_1, which receives the infrared beam by fibre optics cable or via a lens in free-air tranmission. The fibre optics cable should be the

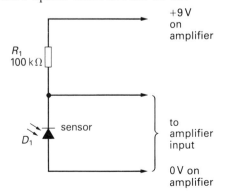

Figure 22.23 Circuits for the Infrared Emitter and Sensor

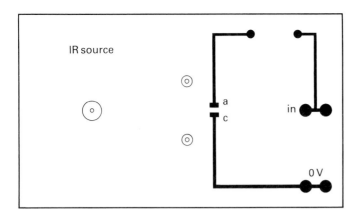

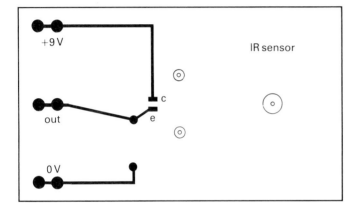

Figure 22.24 PCB track pattern

polymer type, which has an overall diameter of 2.2 mm. The cable ends are cut with a scalpel and trimmed of covering. They are then pushed into the plug assembly.

Components and materials

LED$_1$: infrared light emitting diode type OP160
Tr$_1$: infrared phototransistor, type OP500
device housing: 2-off for LED$_1$ and Tr$_1$
fibre optics plug: 2-off for terminating the ends of fibre optics cable

These are available from Farnell Electronic Components Ltd, Canal Road, Leeds LS12 2TU.

lens: 2-off convex 5/8 inch diameter type 5525 from Combined Optical Industries Ltd, 200 Bath Road, Slough SL1 4DW
tubing: 2-off black PVC, 50 mm long, 5/8 inch diameter, for holding lens press-fitted into one end
saddle clip: 2-off for holding tubing
screws: 6-off for holding device housing (2 each) and saddle clip
PCB: 2-off 90 mm × 50 mm

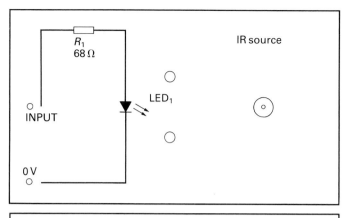

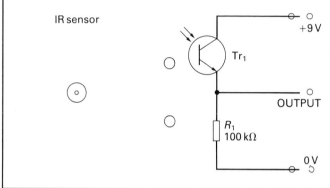

Figure 22.25 Component overlay

R_1: 2-off fixed value resistors, 47 Ω and
   100 kΩ
battery: 9 V PP9
battery clip: PP9 type
wire: multistrand, e.g. 7 × 0.2 mm; single
   strand e.g. 1 × 0.6 mm
connectors: PCB header and PCB socket
   housing; crimp terminals

Using the Infrared Source and Sensor

1   Speech Link
Figure 22.26 (overleaf) shows the system
required for transmitting speech over the
infrared beam. Two Loudspeakers
(Project Module A3), and two Audio
Amplifiers (Project Module A5) are
needed. One Loudspeaker is used as a
microphone connected to the input of
the Audio Amplifier which modulates
the transmitted beam. The second
loudspeaker is connected to the output

of the second Audio Amplifier to
reproduce the speech transmitted. Set
the volume controls on each Audio
Amplifier to maximum gain. The
transmitter and Receiver units will need
to be lined up carefully and set up firmly,
e.g. by using photographer's tripods.

2   Radio Link
Replace the Loudspeaker, which acts as
the microphone in the transmitter, by the
Radio Receiver (Project Module B5) so
that radio programmes can be
transmitted via the infrared beam.

3   Fibre Optics Link
Speech and music can be transmitted
along fibre optics cable by plugging the
cable ends, terminated by fibre optics
plugs, into the device housing on the
sensor and source units. The lenses are
not required but the amplifiers are. Use
fibre optics cables, 2 m long or more, if
required.

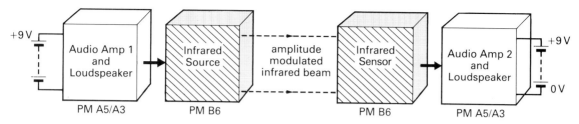

Figure 22.26 System for the speech link

22.8 *Project Module* B7

Metal Detector

What it does

This Project Module detects metallic objects, e.g. buried coins. It is used with the Radio Receiver (Project Module B5) so that change in tone is heard when it 'finds' metal.

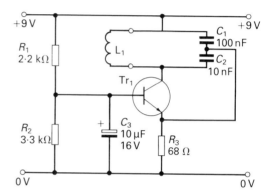

Figure 22.27 Circuit diagram of the Metal Detector

Circuit

Figure 22.27 is a low-power radio transmitter. It produces a carrier wave (an unmodulated radio frequency wave), which has a frequency about the same as that picked up by an AM receiver. This carrier wave interferes (or 'beats') with the carrier wave transmitted by a station on the medium waveband to produce an audible tone. In common with many more sophisticated metal detectors, the frequency of the radio waves produced by this radio transmitter changes if metallic objects are brought near to the coil, L_1. This has the effect of changing the pitch of the note heard from the radio receiver. The pitch increases (or decreases) for a ferrous metal such as iron or nickel, and decreases (or increases) for a non-ferrous metal such as aluminium or gold.

The circuit in figure 22.27 is known as a Colpitts oscillator. It works as an oscillator because of the positive feedback provided

by the connection between the centre-tapped capacitive divider, i.e. C_1 and C_2, and the emitter of transistor Tr_1. Self-sustaining oscillations are generated in the tuned circuit, made up of a search coil L_1, and C_1 and C_2. A fraction of the voltage generated across the tuned circuit is fed back to the emitter to sustain the oscillations.

The search coil is made by winding about 150 turns of 0.315 mm enamelled copper wire on a 100 mm diameter former, e.g. the plastic screw cap from a coffee jar as shown in figure 22.28. Do not use a metal former. Make sure the turns are close wound on the screw cap, and hold the turns in place with PVC tape. Scrape the enamel off the ends of the wire before clamping the ends in the two-way terminal block. Take two long leads from this terminal block and terminate them with PCB sockets so that they can be connected to the PCB pins on the input of the Metal Detector.

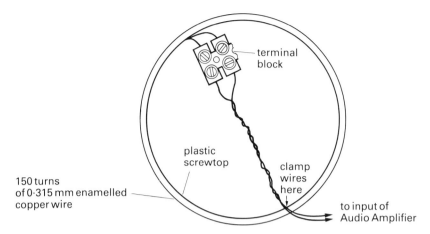

Figure 22.28 The general appearance of the Search Coil

Components and materials

R_1 to R_3: fixed-value resistors, 0.25 W, ±5%

C_1, C_2: polyester capacitors, values 100 nF and 10 nF, respectively

C_3: electrolytic, value 10 μF 16 V working

Tr_1: npn transistor, BC108 or similar

battery: 9 V PP9

battery clip: PP9 type

wire: multistrand, e.g. 7 × 0.2 mm; single strand; e.g. 1 × 0.6 mm

PCB: 90 mm × 50 mm

connectors: PCB header and PCB socket housing; crimp terminals

enamelled copper wire (see text) for L_1 (see text)

PCB assembly

Figure 22.29 (overleaf) shows the layout of the components on the PCB, and figure 22.30 (overleaf) the copper track pattern on the other side of the PCB. See Section 6.3, Book A, for guidance on the preparation of the PCB.

The terminal pins for the connections on the PCB are made by cutting single and double pins from the PCB header. Single sections of the PCB socket housing are used for terminating wire ends for connecting the Metal Detector to the search coil, L_1. Strip 5 mm of insulation from the ends of the wires, and use a crimping tool to squeeze a crimp connector on the bare ends. Push the crimp connector into the PCB socket until it clicks into place.

Testing and use

Connect a 9 V battery to the Metal Detector and plug in the search coil. The circuit should now be producing radio waves. Set up the Radio Receiver (Project Module B5), Audio Amplifier (Project Module A5) and Loudspeaker (Project Module A3), using a separate 9 V battery and tune the radio into a medium wave station. You should hear a 'whistle' as you tune across a station. If you just tune off the station this will reduce the speech or music but the whistle should still be heard. Temporarily disconnect the battery from the Metal Detector to check that the whistle is due to interference between the carrier wave produced by the station and the radio waves produced by the Metal Detector. You may, of course, use a portable radio instead of Project Module B5.

Now bring various bits of metal up to the search coil, L_1, and note the change in pitch of the whistle as the objects pass the coil. Can you tell the difference between ferrous (e.g. iron) and non-ferrous (e.g. copper) objects? If the whistle you hear is not very strong, or if it is too high or low in pitch, you should change the value of R_2.

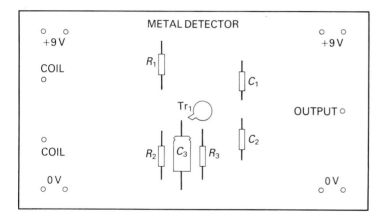

Figure 22.29 Component layout on the PCB (actual size)

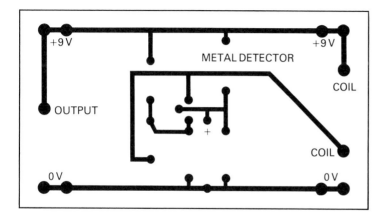

Figure 22.30 Track pattern for the PCB (actual size)

Questions and Answers — Book B

Revision questions

If you have taken the quicker route through the book, do not answer questions marked with an asterisk (*).

General

1 Electrical resistance in a circuit causes electrical energy to be converted into heat.
   True or false?

2 The filament of a lamp has a....resistance, so that it converts electrical energy into heat and light.
   High or low?

3 The filament of a lamp is made of material called...
   tungsten or copper oxide?

4 The coulomb is a unit for measuring the charge on an electron.
   True or false?

5 An electric current is a flow of...
   atoms, electrons or volts?

6 A current of two amperes is caused by electric charge flowing at the rate of....coulombs per second through a conductor.
   One, two or three?

7 The letters e.m.f. mean...
   extra mighty flow, electrical meeting force or electromotive force?

8 What can you say about the current strength at different points in a series circuit?

9 A voltmeter is connected....a component in order to measure the voltage between its terminals.
   in series with, across or between?

Resistors

10 What is wrong with circuit shown in the figure 23.1?

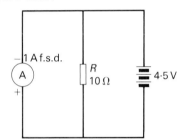

Figure 23.1

11 Resistance = .../amperes.
   volts, ohms or current?

12 A 6 V, 100 mA lamp has a resistance of....ohms.
   100, 600 or 60?

13 One megohm equals a....ohms.
   Thousand, million or hundred?

14 Write 18 000 ohms in kΩ.

15 A resistor has bands coloured in the order orange, white, red, silver. What is its value and tolerance?

16 What would be the colours on a 47 kΩ resistor?

*17 A resistor marked with the British Standard code 4K7G has a value of ...
   47 kΩ ± 10%? / 4.7 kΩ ± 2%? / 47 Ω ± 2%?

*18 Write down the value of a resistor marked as 330 KJ.

*19 The lowest possible value of a 1.2 kΩ, ± 10% resistor is...
        1.08 kΩ? / 1.32 kΩ? / 980 Ω?

20 What is the maximum reading on a 500 mA f.s.d. milliammeter?

21 $R = V/I$ is Ohm's law.
        True or false?

22 What is the resistance of a resistor if a voltage of 4.5 V between its ends causes a current of 1.5 mA to flow through it?

23 A 3.3 kΩ resistor has a current of 2 mA flowing through it. Which voltmeter would be most suitable for measuring the voltage across the resistor?
        50 V f.s.d, 5 V f.s.d. or 10 V f.s.d.?

24 What is the current through a 470 ohm resistor when the voltage between its ends is 2.35 V?

25 A 4.7 kΩ resistor connected in series with an unknown resistor has a total resistance of 10.3 kΩ. What is the value of the unknown resistor?

26 When two resistors are connected in parallel, the overall resistance is....
        higher, lower or unchanged?

27 A 15 kΩ resistor is to have its effective value reduced to 12 kΩ. What is the value of the resistor required to be connected in parallel with it?

28 Calculate the effective resistance between points A and B in figure 23.2.

(i)

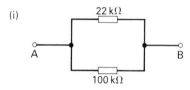

(ii)

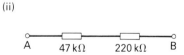

Figure 23.2

(iii)

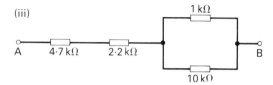

Figure 23.2

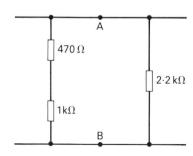

Figure 23.3

29 Calculate the equivalent resistance between A and B of the arrangement shown in figure 23.3.

30 The internal resistance of a battery has the effect of....its terminal voltage when current is drawn from the battery.
        increasing, equalising or decreasing?

31 A 5kΩ resistor has a potential difference of 12 V across it. Calculate (a) the charge which passes through it in 5 minutes and (b) the energy converted into heat in the resistor in 100 seconds.

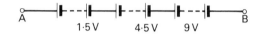

Figure 23.4

32 What is the maximum voltage available across the terminals A and B of the circuit shown in figure 23.4? What is the smallest voltage above zero that you could obtain from this series circuit of batteries?

33   What is the voltage across R_2 in the circuit in figure 23.5?

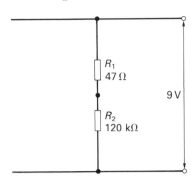

Figure 23.5

Voltage dividers/Wheatstone bridge

34   A voltage greater than the supply voltage can be obtained from a voltage-divider.
    True or false?

35   A voltage-divider has....terminals.
    One, two or three?

36   When a potential-divider is loaded, current is drawn from it by a resistance connected across one part of it.
    True or false?

37   The end-to-end resistance of the voltage-divider shown in figure 23.6 is 2.0 kΩ. It is connected to a 9 V source, and the load is taken from the midpoint of the resistor. Find the load voltage and the current drawn from the source when the load resistance is (a) infinitely large; (b) 9.0 kΩ; (c) 3.0 kΩ.

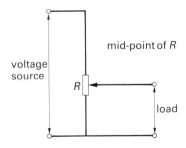

Figure 23.6

38   A Wheatstone bridge is said to be balanced when no current flows through the meter shown in figure 23.7.
    True or false?

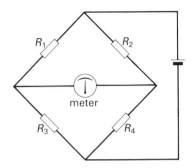

Figure 23.7

39   A Wheatstone bridge measures...
    resistance, current or voltage?

40   In the circuit of question 38, $R_1 = 100$ Ω, $R_2 = 680$ Ω, and $R_3 = 200$ Ω. The value of R_4 which makes the bridge balance is....ohms.
    680, 1500, 1360 or 1280?

Energy and power

41   The joule is a unit of....
    temperature, energy or current?

42   The watt is a unit of....
    energy, power or voltage?

43   A 60 W lamp converts..........joules of electricity per second into light and heat.
    60, 6 or 240?

44   Joules = amperes × volts.
    True or false?

45   A 3.3 kΩ resistor passes a current of 10 mA. What wattage should this resistor have?

*46   A 300 mW bead thermistor has a resistance of 200 ohms and the voltage across its ends is 9 volts. Does the power dissipated in the bead exceed its rated maximum?

47   What do the letters LDR stand for?

48   An LDR has a resistance
which....when it is illuminated.
Increases, varies or decreases?

49   What does the word 'thermistor'
mean? Does it describe the action of
the device?

Capacitors

50   Figure 23.8 shows different types of
capacitors. Answer the following
questions about them.

(a)   What type of capacitor is C_1?

(b)   You need a capacitor of value
between 100 pF and 500 pF in a
circuit where the maximum voltage
across the capacitor will be 240 V.
Which capacitor would you choose?

(c)   You are making a radio and want to
tune into different stations. Which
capacitor would you choose?

(d)   You need an electrolytic capacitor
whose value is greater than 5 μF but
is a low-leakage type. Which
capacitor will be suitable?

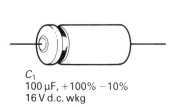

C_1
100 μF, +100% −10%
16 V d.c. wkg

C_4
silvered mica
100 pF, ± 1%.
500 V d.c. wkg

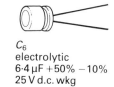

C_6
electrolytic
6·4 μF +50% −10%
25 V d.c. wkg

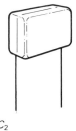

C_2
polyester
0·47 μF, ± 10%
250 V d.c. wkg

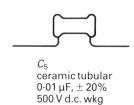

C_5
ceramic tubular
0·01 μF, ± 20%
500 V d.c. wkg

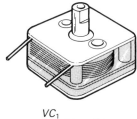

VC_1
variable solid
dielectric
365 pF

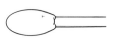

C_3
solid tantalum
15 μF, ± 20%
15 Vd.c. wkg

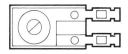

VC_2
trimmer 30 pF
compression
type

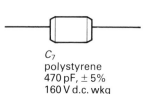

C_7
polystyrene
470 pF, ± 5%
160 V d.c. wkg

Figure 23.8

(e) What is the advantage of the design of the capacitor C_2 compared with the design of C_6?

(f) Which of all the capacitors shown would you expect to have an actual value closest to its marked value?

(g) Which capacitor has the highest d.c. working voltage?

(h) Which electrolytic capacitor would you expect to have the lowest leakage current?

(i) What are the upper and lower values you would expect C_5 to have?

(j) C_4 and C_7 are connected in parallel. What is the value for the equivalent capacitance?

(k) What could be the upper value for C_1?

(l) The trimmer VC_2 is connected across C_4 in a circuit to 'trim' the value of C_4. What would you expect to be the maximum value of the trimmed C_4?

(m) Capacitors C_3 and C_6 are connected together in parallel. What would be their capacitance?

(n) Capacitors C_1 and C_3 are connected in series. What is their equivalent capacitance?

(o) What is the lowest value you would expect for C_6?

(p) Is it possible to obtain a value of 160 μF by combining some of the capacitors?

51 The capacitor shown in figure 23.9 is a 22 μF tantalum type. the multimeter is set to the 'ohms × 10 000' range, and its leads are connected as shown. What do you expect to see the multimeter register?

52 What is the charge stored in a 22 μF capacitor when it is connected to a 9 V battery?

53 How would you show that a capacitor stored electrical energy?

54 A 470 pF capacitor is connected in parallel with a 150 pF capacitor. What is the total capacitance of this combination?

55 A 100 nF capacitor is connected in parallel with a 4700 pF capacitor. What is the total capacitance in nanofarads and microfarads?

56 A 30 μF and a 6 μF capacitor are connected in series. What is the value of the total resultant capacitor?

57 Three capacitors of 10 μF, 8 μF and 40 μF are connected in series. What is the value of the total resultant capacitor?

58 Find the quantity of charge in coulombs stored on a small 200 pF capacitor charged to 10 kV.

59 A large 40 μF capacitor has been charged up to 150 V. What quantity of electrons in coulombs is stored on this capacitor?

60 Two capacitors $C_1 = 50$ μF and $C_2 = 10$ μF are connected in parallel to a d.c. voltage of 12 V. Find the charge on each capacitor.

61 Two capacitors $C_1 = 100$ pF and $C_2 = 25$ pF are connected in series across a 10 V supply. What is the charge on each capacitor and the voltage across each?

Time constant

62 What is the period of time for one time constant if a 220 μF capacitor is connected in series with a 22 kΩ resistor?

63 A 0.005 μF capacitor has been charged to 200 V. If it is permitted to discharge through a 50 kΩ resistor,

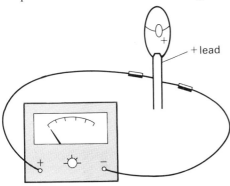

+ lead

Figure 23.9

find the initial current which flows. What is the voltage across the resistor and across the capacitor at this instant?

64 Calculate the time constant of a circuit in which a 10 μF capacitor is in series with a 100 kΩ resistor. What is the effect on the time-constant of increasing the value of the resistance?

65 Calculate the charge and the energy stored by a 50 μF capacitor if a potential difference of 20 V is applied across it.

66 What is the time constant of a 100 μF capacitor and a 47 kΩ resistor? If a p.d. of 10 V is applied across the resistor and capacitor, how long will it take for the voltage across the capacitor to reach 6.3 V?

67 Explain the general shape of the charge and discharge voltage-time characteristics for a capacitor in series with a resistor.

Reactance

*68 What is meant by the reactance of a capacitor? Draw a graph showing how the reactance of a capacitor varies with frequency.

*69 Write down an equation for the reactance of a capacitor.

*70 It is known that in a.c. circuits a capacitor offers a resistance to the current flowing in the circuit. Why do capacitors not have a resistance value marked on them?

*71 Which of the meters in the circuit shown in figure 23.10 will register the higher current?

*72 Calculate the reactance of each of the two capacitors in the circuit for question 71.

*73 Assuming that the supply voltage, 20 V, is the r.m.s. voltage, calculate the r.m.s. current which flows

through each capacitor in the above diagram. Ignore the resistance of the milliammeters.

*74 What is the reactance of a 100 nF capacitor at frequencies of 10, 50, 500 and 1000 Hz?

*75 Design a circuit using a 555 timer which will energise a relay for 10 seconds periods at intervals of 3 seconds. What is the general name for the circuit you have designed?

*76 Design a general-purpose timer based on a 555 which will energise a relay for selected periods of 1 minute, 4 minutes and 15 minutes after pressing a button.

Inductors

77 An inductor does not conduct alternating current.
      True or false?

78 An inductor is sometimes called a 'choke', because it....the passage of alternating current through it.
      resists / multiplies / stops?

79 The inductance of an iron-cored inductor is increased if the core is removed from the coil.
      True or false?

80 Inductance is measured in units of...
      farads / ohms / henrys?

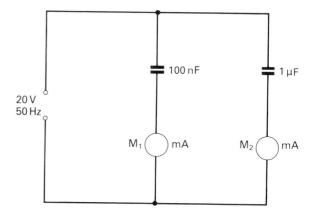

Figure 23.10

81 When a steady current flows through a coil of wire, the coil behaves as if it were a...
      magnet / resistor / inductor?

82 A changing magnetic field is said to induce a back....in the conductor carrying the current producing the field.
      current / inductance / electromotive force?

83 Self-inductance is caused by a changing current in a coil producing a changing magnetic field which induces an e.m.f. opposing the changing current.
      True or false?

*84 Write down the equation for the time constant of an *L-R* series circuit.

*85 Calculate the time constant of a circuit consisting of a 2 H inductor in series with a 100 Ω resistor.

*86 The resistance to current flow which an inductor presents in an a.c. circuit is known as...
      capacitance /
      resistive impedance /
      inductive reactance?

87 What does a transformer do?

88 Draw the electronic symbol for a transformer having a laminated iron core.

89 A transformer core is laminated in order that the transformer may more efficiently convert....energy from one voltage to another.
      heat / electrical / sound?

90 What are eddy currents?

91 A mains-to-low-voltage transformer converts 240 V to 12 V a.c. What is the turns ratio of the windings of this transformer? If there are 120 secondary turns, how many turns are there on the primary coil?

92 What does a step-up transformer do?

93 Name two types of transformer core materials.

94 An electromagnetic relay enables a small current to switch a large current.
      True or false?

95 Name and draw three types of contacts of a simple relay.

96 Why do the armature and core of a relay have to be made of a magnetic material called soft iron?

Revision answers

6   Two

12   60 ohms

13   Million

14   18 kΩ

15   3.9 kΩ, ± 10%

17   4.7 kΩ, ± 2%

18   330 kΩ, ± 5%

19   1.08 kΩ

20   500 mA

22   3 kΩ

23   10 V f.s.d.

24   5 mA

25   5.6 kΩ

27   60 kΩ

28   (i) 18 kΩ: (ii) 267 kΩ; (iii) 7.8 kΩ

29   0.88 kΩ

31   (a) 0.72 C (b) 2.9 J

32   15 V; 1.5 V

33   6.5 V

37   (a) 4.5 V and 4.5 mA; (b) 4.3 V and 4.7 mA; (c) 0.43 V and 5.1 mA

40   1360 Ω

43   60 J/s

45   0.5 W

50   (i) 0.012 µF and 0.008 µF respectively
(j) 570 pF
(k) 200 µF
(m) 21.4 µF
(n) 13 µF
(o) 5.8 µF

52 2.0×10^{-4}C

54   620 pF

55   147 nF or 0.147 µF

56   5 µF

57   4 µF

58   2 µC

59   6 mC

60   600 µC (50 µF); 120 µC (10 µF)

61 2×10^{-10} C; 8 V (25 pF), 2 V (100 pF)

62   5 s (approximately).

63   4 mA; 200 V.

64   1 s.

65   1 mC; 0.02 J.

66   4.7 s (approximately); 4.7 s

72   32 kΩ (100 nF); 3.2 kΩ (1µF)

73   0.37 mA (100 nF); 3.7 mA (1 µF)

74   160 kΩ; 32 kΩ; 3.2 kΩ; 1.6 kΩ

85   20 ms

91   240/12 or 20/1; 2400

Answers to questions

Section 2.2
2   4 C/s
3   0.3 C/s

Section 2.5
1   3 Ω
2   960 Ω
3   100 Ω

Section 3.6
1   12 kΩ; 220 kΩ; 3 MΩ
2   100 kΩ
3   0.47 kΩ
4   1000
7   A — 56 kΩ, ± 10%; B — 2.2 kΩ, ± 2%;
C — 33 Ω, ± 5%

Section 3.7
1   1.2 MΩ ± 1%; 150 Ω ± 20%; 12 kΩ ± 5%; 68 kΩ ± 10%
2   330 KK; 47 RJ

Section 3.8
1   110 Ω
2   100 Ω; 33 Ω; 680 Ω; 1.5 MΩ

Section 4.2
1 7 Ω (approximately)
2 1.2 mA
3 13 V
4 50 Ω

Section 4.3
3 2.0 kΩ
4 6.9 kΩ; 6.8 kΩ

Section 4.4
3 2.35 kΩ; 2.2 kΩ

Section 4.5
1 8.1 kΩ
2 77 kΩ
3 11.8 kΩ

Section 4.7
3 8 Ω

Section 5.1
1 (a) 4 V; (b) 3 V; (c) 1.67 V;
     (d) 6.67 V; (e) 2.2 V

Section 5.2
1 0.4 V
2 About 93%
3 About 4%

Section 5.3
1 454 Ω

Section 6.3
1 300 J
2 20 h

Section 6.4
2 1000 W (1 kW)
3 21 W
4 67 mA; 3.6 kΩ
5 100 mW; 7.2 V

Section 6.5
1 10 V; 0.01 W
2 Yes

Section 6.6
1 1.3 kΩ (1.2 kΩ preferred value)
2 500 Ω (470 Ω preferred value)

Section 6.7
1 25 C
2 9 V/m
3 (a) 20 mA; (b) 1.8 N; (c) 18 J

Section 7.3
2 1 MΩ, at least

Section 8.1
2 0.9 C

Section 8.2
1 10 μF; 400 μF; 1000 μF; 5 nF; 3.3 pF

Section 8.4
1 99 pF; 101 pF
2 70 μF and 60 μF

Section 10.2
1 100 kΩ; 1.4×10^{-2} s; 200 μF; 30 s;
     1 MΩ

Section 11.6
1 4000 μF

Section 12.1
1 10 000

Section 12.2
1 (a) 0.2 C; (b) 300 μC; (c) 0.125 μC
2 1.5 mC (millicoulombs)

Section 12.3
1 0.16 J
2 10 V; 0.005 J

Section 13.3
1 0 A

Section 13.4
1 160 kΩ; 32 kΩ; 3.2 kΩ; 1.6 kΩ;
2 (b) 1 kHz (c) 1.6 kΩ (d) 100 pF; (e)
     1 kHz (f) 1 kHz (g) 1 kHz (h) 10 Hz (j)
     1 μF (k) 16 kΩ (l) 100 pF

Section 17.4
1 one thousandth of a henry
2 one millionth of a henry
3 3 mH; 5 μH; 20 mH

Section 18.2
1   0.4 ms
2   2 μs

Section 18.14
3   6.28 Ω; 125.6 Ω
4   1.6 H

Section 19.4
1   Smaller secondary — 6 V a.c.
    Larger secondary — 42 V a.c.
3   40 turn winding — 8 A
    280 turn winding — 1.14 A